THE 10 BIG QUESTIONS ABOUT THE FUTURE

Virtual Conversations with Past Masters of Innovation

dedicated to Sara and Lorenzo...

... they are starting now to ask the big questions.

INDEX

emerge to improve diagnosis, treatment and disease prevention?

The 10 big questions about the future

"The present is theirs; the future, for which I really worked, is mine." -
Nikola Tesla

Preface

Andrea Ferrante and I met on 31 December 2099. It was hot at the time, much hotter than the summer of far-off 2023 when we complained about the crazy temperatures ('the hottest July ever!'), when we did not yet know that it would be one of the 'coolest' seasons of the following decades. Andrea and I used to travel through time, specifically into the future (but not only). We did it for work and for passion. Perhaps today we can also say that we travelled out of necessity: to explore, to investigate, to understand, to find. Travelling into the future has become, shall we say, compulsory since we have stopped, as human beings, asking ourselves prospective questions in order to scrutinise the horizons of tomorrow: what will happen? What awaits us? In the first decades of the 21st century, we were so crushed by contingency (emergency?), so focused on obtaining immediate results (returns?), that we had somewhat forgotten how to interrogate the future, in search of narratives of possible futures.

Out of metaphors, with Andrea we shared an approach to innovation, work and, ultimately, life, nurtured by the 'necessary curiosity' to question the consequences of the present (and the past). Foresight methods, futures studies, strategic foresight, are a set of systematic techniques that make it possible to explore alternative, long-term future scenarios and support the design and implementation of strategic paths in the short term. In order to do so, it is particularly necessary to activate a type of future-oriented mindset, or rather possible futures, through systemic looking and thinking. This makes it possible to observe the phenomena and trends of the present (and the past) through the lens of the possible future implications they will have. Such a scanning of what is happening today nurtures fundamental attitudes and skills: curiosity, imagination, creativity, vision. One becomes good at picking up signals, at first very weak, that may have future developments (becoming 'emerging phenomena') and trigger change processes over the years ('trends') and transformative macro-forces over the decades ('megatrends') that will produce impacts in different fields, sectors, contexts, latitudes.

Andrea does this job very well: his podcast 'The Future Of' has for many years been synthesising a multiplicity of these observable signals around us, wondering and wondering what else they might become. The future not as an answer (to be unveiled), but as a question: the future to be 'made possible', we say. So if there is not just one future, but numerous, alternative ones, then in that range of possible scenarios there will also be one that corresponds to the wishes of a community, expressing what is socially, economically, environmentally, politically preferable. Technologies are a perfect tool for reflecting on implications and exploring possibilities. Therefore, this essay by Andrea plays a decisive role in helping us to think: what are the evolutions of AI, digitisation, virtual reality, etc.? And what effects on work, education, health, lifestyles, social models?

Alberto Robiati

Introduction

You deal with the future even without knowing it. Always have been. There is no need to be futurists, analysts, thinkers or visionaries. There are questions about the meaning of our lives that we have all asked ourselves at one time or another. Where do we come from? Did the big bang really happen and is the universe expanding? Are we alone in the Universe? Will life continue on this planet or will we become extinct? What is there after death? Does our life have any deep meaning or is it just a collection of chemical reactions?

These questions, laden with an intrinsic curiosity, have their roots in the human soul since ancient times. Our thirst for knowledge, inherent in our being, constantly drives us to seek answers to such questions, to peer into the darkness of the unknown in search of clues, so that we can weave the fabric of our understanding of the world and our place in it.

And when we ask ourselves these questions, it is because we have the desire to peek through the keyhole of the door of infinity and see what awaits us in the future. Even explanations of past events, in their own way, express our orientation towards tomorrow. To understand what has happened, to hypothesise what will happen.

What about our future? It is crucial, as it weaves the threads of our collective responsibility. We wonder whether life on our planet will continue to thrive or whether, in a dark and distant future, the traces of our existence will only be fossilised scars in geological history. Science and technology, as well as our ethical choices, will be the weavers of this future web, determining whether we leave a legacy of prosperity or destruction.

I will not be the one to give you the crucial answers to the above questions, and that alone might be a good reason to abandon reading right away. However, even though I cannot offer definitive answers to these existential questions, I invite each of you to continue searching, exploring and contemplating. The answers may change with time, but the search itself is a fundamental element of the human condition.

Whether you are driven by the instinct of discovery, the flame of knowledge or the desire for meaning, remember that in seeking the answers to these questions, you are in effect charting the very course of your humanity.

I just have a tiny observatory on the future, focused on technology. I am the author of the podcast The Future Of, which has a very simple and powerful slogan "the space of the future-curious". I look at the future through the lens of technology, and have done so for many years. After several hundred episodes, studies of future exploration methods, corporate speaking engagements and participation in future communities of practice, I can say that it's a bit like opening the eye of an astronomical observatory and pointing at a piece of the sky. Some focus on the moon, some on planets, some look for black holes, some want to grasp the immensity of the Milky Way.

I focus on technologies. Which are one of the most important drivers of human innovation, but certainly not the only one. There is innovation in the social field, in economics, in philosophical thinking, in politics and so on. I have chosen my point of view. I make no claim that it is better than another, indeed the juiciest fruits in thinking about the future come precisely from the integration of different disciplines. As is often the case, diversity is an asset: you will soon understand why this concept is also applied in this book.

Just like the questions that the common man asks about the meaning of things, I have distilled 10 questions about the meaning of technology and its role. I think they are fundamental questions. I am convinced that trying to answer them can be a good starting point to speculate on the future, open a good debate and aspire to the future. Because the task of a futurist is not to predict the future, but to yearn for the future, to be ready for different scenarios, to help one's neighbour and oneself to direct what will happen. In our lives, in those of our work organisations and in those of the communities in which we participate.

These are all questions that foreshadow possible scenarios that will have a direct impact on our lives in the next 10-20 years. There is no doubt that we would be happy to know that a few million light years away we have found intelligent life, but once the revelry of discovery is over, you would probably go back to your everyday life. And the same would happen if someone told you with irrefutable certainty that there is nothing after death. Or a certain percentage probability that you would be reincarnated as a mouse. If then, instead of the big bang, some scientist revealed on the news that the source of everything is a white hole spewing out matter caught in some elsewhere we don't know about, I'm pretty sure your breakfast - office - gym - aperitif - Netflix routine wouldn't change much.

On the contrary, if someone were to explain to you that in less than 10 years your job will disappear and be replaced by a very clever artificial intelligence (or rather, the clever one would be the one who owns it), I don't think you would sleep soundly. From the next day, you would start thinking about how to 'recycle' yourself before you become useless.

I dare not imagine how you would react to knowing that in a few years your beach house, lying along a coastline that will be submerged by rising seas, will no longer be worth anything. Assuming it doesn't go completely underwater. You would probably call your estate agent to give him a mandate to sell.

And what if they presented you with the scenario that your children will only study in virtual classrooms, with high-powered visors, confronting multilingual people from afar. On top of that, learning professions, which will probably already have died out before your kids even get a degree or a very expensive MBA. Wouldn't you immediately rush to reflect on how to orient your and their next decisions?

The 10 questions

All futures are important, but those that have technological change as a driver may be more imminent and impactful. If you are far-sighted enough to think that what may happen in a decade or so requires thinking about already today, here are the 10 issues of paramount importance that I bring to your attention.

1. What will be the economic and social impacts of artificial intelligence? How will AI affect employment, the distribution of economic resources and the social structure?

2. How to ensure data security and privacy in the digital age? What ethical and legal challenges arise from the widespread use of technology and the massive collection of personal information?

3. What will be the environmental impacts of accelerated technological innovation? How to balance technological development with environmental sustainability and the needs of the planet?

4. How will educational models and learning change with the ever deeper integration of technology in classrooms and online training? What skills will be relevant in the digital future?

5. What ethical implications arise from the use of technologies such as biotechnology and advanced genetics? How can fundamental human values be preserved while exploring new scientific frontiers?

6. What are the risks and benefits of advanced automation, including vehicle and industrial process automation? How to balance efficiency with the preservation of jobs and safety?

7. How will artificial intelligence and robotics affect the health and healthcare sector? What opportunities emerge to improve diagnosis, treatment and disease prevention?

8. What are the cultural implications of virtual and augmented reality? How will our social interactions, art forms and human experience change in the context of immersive digital worlds?

9. Will we be able to visit other planets far from Earth? What great surprises await us? Are we alone in the Cosmos?

10. What will be the impacts on the distribution of global resources when technology could widen the gap between developed and developing countries? How to ensure equitable access to technology opportunities worldwide?

These questions reflect some of the central concerns and challenges humanity is facing as it moves towards an increasingly technological and interconnected future.

You might object that there is no need to ask questions. We can live without them. Head down and pedal. But let me plead the cause of those who like to ask questions (and try to give answers). In the vast landscape of human enquiry, questions have always been the compass that guides us through the uncharted territories of knowledge and understanding. They are the spark that ignites curiosity, the chisel that carves the paths to new insights and the driving force behind every great leap of progress. In the realm of contemplation of the future, questions become more than mere tools; they evolve into beacons that illuminate the road ahead.

When we stand at the crossroads of time, peering into the unknown that awaits us, it is the questions we dare to ask that propel us forward. The very act of asking questions challenges us to examine our assumptions, to transcend the boundaries of our current understanding and to imagine possibilities that extend beyond the horizon. It is precisely through questions that we are able to unravel complexity, discern patterns and unearth the hidden gems of foresight.

Consider the innovators and visionaries who reshaped history. Their research into what could be, rather than what already is, fuelled their imaginations and stimulated their audacity. It is with their questions that they harnessed the winds of change, guiding humanity into uncharted waters where innovation thrives.

In the following pages, we delve into ten big questions, ten monumental questions about the future. These questions are not intended to provide definitive answers, but rather are gateways to realms of exploration. Each question is a catalyst, a starting point for contemplation and dialogue. They invite us to step out of our comfort zone, to challenge established norms and to construct the set of possibilities that tomorrow holds.

The power of these questions lies not only in their ability to provoke thought, but also in their ability to foster connection. When we engage with these questions, we engage with others across time and space. We engage with the minds of the past, present and future. Through the conversations that emerge, we transcend generations, ideologies and cultures. We participate in a timeless dialogue about the human condition and the ever-changing nature of progress.

The answers

So I hypothesised answers, and herein lies the surprise, I did not do it myself. After all, what could most people care about the opinion of an Andrea Ferrante? But if it were Albert Einstein, Steve Jobs, Stephen Hawkings, Nikola Tesla, Marie Curie, Sigmund Freud and many other incredible men and women of science and innovation who spoke, that would be a different story.

In the grand picture of human history, the brushstrokes of innovation have been painted by extraordinary individuals whose ideas transcend their time, leaving an indelible mark on the path of progress. As we embark on this exploration of the future, we find ourselves not only looking forward but also back, drawing inspiration from the luminaries who paved the way with their bold thinking and revolutionary creations. The idea of engaging in virtual conversations with these past masters of innovation is a portal to understanding the very essence of innovation. It is a journey of rediscovery, where the wisdom of the past meets the challenges of today and the opportunities of tomorrow.

What would Einstein think of artificial intelligence? What would Alan Turing say about the digital age and privacy? Wouldn't it be fascinating to know what Galileo Galilei would have to say about the environmental impact of technological innovation? And who knows how much the genius Richard Feynman would amaze us by telling us about new models of digital education! And if we really want to get serious, wouldn't it be borderline provocative to talk about automation with Henry Ford and Thomas Alva Edison? Don't you think that Asimov and Verne might have something interesting to say about virtual reality and digital worlds? Perhaps in the form of a short story!

You are about to embark on a journey through time and into the minds of some of history's greatest thinkers, innovators and visionaries. Men and women who, despite coming from different eras and unique backgrounds, shared a burning curiosity for the unknown, an unwavering passion for innovation and a determination to shape the course of the future. Although their lives were marked by challenges, obstacles and revolutions, their spirit remains an undying source of inspiration, capable of provoking deep reflection and crucial questions about technology and its impact on our existence.

By participating in these virtual conversations, we extend an invitation to the minds that have dared to dream, the thinkers who have dared to challenge convention and the visionaries who have dared to see beyond the ordinary. Their stories and ideas remind us that innovation is not just about gadgets and gizmos; it is about reshaping perspectives, challenging norms and forging new paths.

When we embrace the perspectives of these masters of innovation, we begin to recognise patterns and principles that cross generations. Their struggles, triumphs and reflections become a beacon that guides us through the fog of uncertainty surrounding the future. Through their virtual voices, we find echoes of our own challenges and aspirations, discovering that the human quest for progress is a continuum, a relay race in which the baton of inspiration is passed from one generation to the next.

In this book, I have curated a collection of conversations that cross time, joining us with the genius of minds that have long since left the physical realm but continue to live on through their ideas. Their insights become lenses through which we view our own endeavours, refining our understanding of the evolving nature of innovation. And, lest we never forget, in these dialogues we learn that innovation is not a solitary endeavour but a collective symphony of minds spanning the centuries.

If reading these last lines you found yourself nodding or, better still, foaming at the mouth like Pavlov's dogs, or if you felt the same joy as when you solved Fermat's theorem... then this book is for you.

A recommendation

I encourage you to think that this journey is not passive. I invite you to be an active participant, a co-investigator in the story of exploration. As you navigate these conversations, allow your mind to wander freely, ask bold questions and make connections that resonate with your own experiences. Innovation is not confined to the past, nor is it reserved for the elite; it is a calling for every curious soul willing to venture beyond the boundaries of the known.

Many future exercises begin by putting experiences, visions, reflections and even prejudices on the table. Those who want to build scenarios must let out what they have inside. This is not a novel; on the last page you will not discover the killer. If anything, you might discover that the 10 fundamental questions might be followed by 10 equally important ones. Or that each of them, in turn, could be broken down into a further 10 sub-questions, expanding your gaze to the infinite. Or to the microscopic, depending on the direction taken.

I encourage you, therefore, to engage with the contents of these pages not as a detached observer, but as an active seeker of knowledge. Stop and reflect on the insights shared by the masters of innovation and consider how their wisdom can be applied to your own ventures. Let their stories inspire your own stories of curiosity, discovery and innovation.

Every question you encounter is an invitation to explore the complexity of the future and your role in it. As you read, allow yourself to embrace uncertainty, because it is in the questions, doubts and uncharted territories that the real magic of innovation lies.

You are not alone in this journey. You are part of a community of thinkers, dreamers and activists who share a common goal: to imagine a better tomorrow. So, as you flip through the pages and immerse yourself in these virtual conversations, remember that you are contributing to the ongoing dialogue about human progress. Your thoughts, reflections and insights are part of the fabric of this exploration. If you would like to share your insights with me, you are free to do so, indeed I invite you to do so. I pledge to respond to any emails, comments, provocations.

May this book serve as a compass to guide you through the labyrinth of the future. May it inspire you to ask questions, connect and contribute your unique perspective to the great film of innovation. The journey ahead is one of collaboration, imagination and the relentless quest to understand what lies beyond the horizon. Your journey begins now and the destination is limited only by the boundaries of your curiosity.

1. What will be the economic and social impacts of artificial intelligence? How will AI affect employment, the distribution of economic resources and the social structure?

"Science and technology must serve humanity and improve the quality of life, not generate further inequalities." - Rita Levi Montalcini

Albert Einstein, known for his revolutionary scientific theories, can contribute to the understanding of the economic and social impacts of artificial intelligence (AI) due to his skill in abstract thinking. His theories can inspire analysis of the theoretical foundations of AI and its possible scientific and economic applications.

Steve Jobs, founder of Apple and a pioneer in technological innovation, can offer perspectives on the economic and social transformation related to AI. His contribution focuses on how AI can revolutionise existing industries and create new markets, influencing employment and the distribution of economic resources.

Karl Marx, philosopher and economist, can help examine the impacts of AI on the distribution of economic resources and social structure. His theories on class dynamics and the struggle for control of resources can be applied to assess how AI will affect economic and social inequality.

Rita Levi Montalcini, neurologist and Nobel Prize winner for medicine, can help explore the medical and scientific implications of AI. Her research on nerve cells may be useful in understanding how AI will influence the healthcare sector and public health, with significant economic and social consequences.

Sigmund Freud, founder of psychoanalysis, can help examine how AI will influence the psychological and behavioural aspect of society. His theories can be applied to understand how AI will influence human behaviour, mental health and social dynamics.

Charles Darwin, known for the theory of evolution, offers an evolutionary perspective to explore how AI will evolve over time, adapting to societal needs and influencing social structure and employment. His perspective can help consider how AI will adapt to evolving human needs.

In conclusion, each of these thinkers makes a unique contribution to examining the economic and social impacts of artificial intelligence.

Their wide range of knowledge and theories can guide thinking about how AI will affect employment, the distribution of economic resources and the social structure, better preparing society to meet the challenges and exploit the opportunities of this emerging technology.

Dear friend,

As we contemplate the economic and social impacts of artificial intelligence, we find ourselves at an epochal crossroads. This invention of human ingenuity is slowly rewriting the pages of history, opening up new horizons and presenting unprecedented challenges. The interconnectivity of economic and social processes, intricately intertwined with intricate dynamics, requires a deep understanding of the consequences that AI will bring.

AI, with its capacity for analysis and learning, will dictate the fate of human employment in an unprecedented way. The industrial revolution, which shaped the face of society in past eras, will seem almost prologue to the impact AI will have on employment. The automation of manual and cognitive activities will inevitably give rise to a redefinition of the very concept of human labour.

In this scenario, the distribution of economic resources could be profoundly reshaped. The rise of AI could widen the gap between those who master the skills needed to coexist harmoniously with intelligent machines and those who, for various reasons, remain marginalised by this digital revolution. However, this scenario is not written in stone; it is in our hands to act with wisdom and compassion to prevent wealth from becoming a privilege for the few.

Reflecting on the social structure, I see the foundations of the entire society shaking under the influence of AI. The paradigms of work, education and governance will have to adapt to this new era. Traditional models will be challenged and reworked, putting our adaptability to the test. The potential for social dislocation is something that cannot be ignored; equity and accessibility must guide every step we take.

As we reflect on the future of AI, it is imperative to consider the historical perspective. Every technological advancement, even the most monumental, has had unforeseen and sometimes surprising consequences. This moment, in which the boundaries of artificial intelligence merge with the fabric of our existence, must be approached with an open mind and a critical mind. We cannot embrace innovation without considering the price we might pay.

That said, history has taught us that humanity has an extraordinary capacity to adapt. We have overcome seemingly insurmountable challenges, and this new era will be no different. AI could free us from many tedious tasks and allow us to focus on more creative and meaningful aspects of human existence. It is our duty, however, to ensure that the benefits are shared equally.

In conclusion, the economic and social implications of artificial intelligence are an enigmatic symphony of potential and uncertainty. We are in the process of writing a new chapter in the human saga, and the direction we take will depend on our ability to deal wisely, ethically and understandingly with the challenges that AI presents us. We can face this future with determination, but also with the confidence that together we can shape a world in which AI acts as a beacon of progress, fairness and humanity.

With admiration for the human potential and curiosity that guides our path,

Albert Einstein

STEVE JOBS (1955 - 2011) - Entrepreneur and co-founder of Apple Inc.

Hey there,

When it comes to the economic and social impacts of artificial intelligence, we find ourselves exploring fascinating and ever-changing territory. It's a bit like going back to the 1980s when we were trying to imagine what it would mean to have a personal computer in every home. And now, look how far we've come.

AI, my friend, is going to have a significant impact on employment, the distribution of economic resources and even the social structure. I think back to when we created the first iPhone. Many thought it would be just another phone, but it revolutionised communication, business and our very perception of technology. Here, AI will do the same, if not more.

As far as employment is concerned, there will be changes. AI will automate some tasks, yes, but this does not mean that people will be replaced. Rather, I think it will free up time and energy to focus on more creative, innovative and meaningful tasks. We have seen this with automation in the manufacturing industry. When machines started doing repetitive tasks, people were able to focus on more high-value tasks.

The distribution of economic resources is another intriguing aspect. Look, AI could increase efficiency and productivity in various sectors. This could lead to a greater abundance of resources. But the key here is how to distribute them fairly. We do not want AI to create an even greater division between rich and poor. We must be proactive in ensuring that the benefits are shared fairly and that no one is left behind.

And then there is the social structure. AI has the potential to change the very structure of society. Think of how Apple changed the concept of the smartphone. Before, it was just a thing for business people. Now, it is an extension of ourselves. AI could do the same. It will change our interaction with technology and with others. It could challenge traditional paradigms of work and education. We must be ready to embrace this evolution and create an environment in which everyone can thrive.

But you know, the most important thing to remember is that we are the creators of our future. AI is not just something that happens to us; we are able to shape and guide it. It's like when we launched the first Macintosh. We had a vision and a mission: to put the power of computers in the hands of ordinary people. With AI, we can have a similar vision. We can use it to improve our lives, solve complex problems and make the world a better place.

Ultimately, man, AI will be a transformative force. It will lead us towards a future of extraordinary opportunities and unique challenges. But as I have always said, 'Stay hungry, stay foolish'. Stay hungry for knowledge, innovation and possibility. And don't be afraid to be a little crazy, to experiment and to create something extraordinary. The future is in your hands, just as it was in mine.

See you soon,

Steve

KARL MARX (1818 - 1883) - Philosopher, economist and political theorist

Dear comrade,

When you ask me about the economic and social impacts of artificial intelligence, I am addressing a topic that is intrinsically linked to my critical view of economics and society. Artificial intelligence, as a tool and manifestation of advanced capitalist production, paves the way for profound transformations in the socio-economic fabric. It is important to analyse this question in the light of the fundamental concepts of political economy and class struggle analysis.

Artificial intelligence, in its current form, is a creation of advanced industrial capital. Capitalist enterprises are investing huge resources in the development and implementation of AI in order to increase productivity and maximise profits. This race for automation through AI has the potential to radically transform employment. As I argued in my work 'Capital,' capital constantly seeks to reduce labour costs and exploit human labour more and more intensively. AI-powered automation can lead to structural unemployment as capital seeks to replace human labour with machine labour to increase profit accumulation.

The distribution of economic resources in a context of increasing AI automation raises a crucial question of social justice. In the capitalist system, wealth is unequally distributed and automation could further

23

accentuate this inequality. Capital has always sought to concentrate wealth in the hands of a few, leaving the majority of the population without the necessary means for a decent life. AI, in the hands of capitalist elites, could be used to increase the control and centralisation of wealth, exacerbating already existing class divisions.

Social structure, dear friend, is closely linked to economic organisation. AI can strengthen the powers of the ruling classes and consolidate their positions of privilege. In the capitalist system, social relations are mediated by private ownership of the means of production. AI, in the hands of capitalists, could perpetuate this division between those who hold capital and those who must sell themselves as labour-power. This could profoundly influence society by maintaining a class-based hierarchy.

However, we must not forget that AI itself is neither good nor bad. It is the purpose and context in which it is used that determine its impact on society. In a socialist system, where the means of production are collectively owned and power is more equally distributed, AI could be used for the common good. It could help free people from alienating jobs and enable greater involvement in the planning and organisation of production.

In my analysis, the foundation of radical transformation lies in the class struggle. The working masses, if aware of their condition and potential for change, can act to redistribute power and wealth. The question of AI requires a deep critique of the capitalist system and a reformulation of production relations. AI could be used to automate heavy work and redistribute work time more equitably, enabling greater participation of people in the decisions that affect their lives.

In conclusion, dear comrade, artificial intelligence represents a crucial challenge that requires critical analysis based on the fundamental laws of political economy. In the context of capitalism, it could accentuate inequalities and reinforce existing power structures. However, the potential for change and transformation lies in the hands of the working masses, who can fight for a socialist system in which AI is utilised for the common good and human liberation.

With faith in the power of the working masses and the opportunity for radical change,

Karl Marx

RITA LEVI MONTALCINI (1909 - 2012) - Neurologist and scientist

Dear Reader,

My life has been dedicated to the study of the nervous system and its complex interactions with the organism and its environment. Therefore, approaching the analysis of the economic and social impacts of artificial intelligence from a neuroscientific perspective is a task I embrace with zeal and determination. Artificial intelligence, with its capacity for learning and adaptation, is a field that resonates with my commitment to understanding the structure and functioning of the human brain.

Let us start with employment, an aspect that has always attracted my attention and determination. AI, with its ability to perform complex tasks and learn from vast and diverse data, will undoubtedly have an impact on employment. As a scientist, I am used to carefully examining the available data and evidence. In this case, the data suggest that some jobs could be automated, especially repetitive and rule-based jobs. However, this need not translate into a threat to human employment. This is where my determination comes in.

As a neurologist, I have studied the adaptability of the human brain, its ability to react to new challenges and to develop new connections. Similarly, mankind has proven to be able to adapt to technological changes. AI could create new job opportunities, requiring human skills that go beyond mere automation. Focusing on developing unique skills, such as creativity, problem-solving and empathy, could be crucial to preserving human employment.

The distribution of economic resources is a subject I approach with a critical eye and an analytical mind. As a scientist, I have always tried to understand the complex dynamics that influence the world around us. AI, if implemented wisely, could increase the efficiency of production and improve the distribution of resources. However, it is crucial to consider the ethical aspect of this distribution. As said by a determined mind, social justice must remain at the centre of decisions. We must prevent AI from accentuating existing economic inequalities. A fair and responsible approach could help mitigate these negative effects.

Finally, we explore the influence of AI on the social structure, a topic that has pushed me to tackle complex challenges in my own career. As a neurologist, I have studied the human brain in the context of the individual and society. AI could influence the social structure as it will change the way we interact with technology and each other. However, we must ensure that these transformations are positive and inclusive. A determined mind is not content with superficial changes. We must consider how AI can promote collaboration, knowledge sharing and participation of all sections of society.

In conclusion, dear reader, addressing the economic and social impacts of AI requires an analytical mind, an empathetic heart and unwavering determination. As a neurologist and scientist, I have learnt that challenges can be met through the intelligent use of resources, the pursuit of social justice and the promotion of fundamental human values. AI is a powerful tool, but we must remember that it is in the hands of humanity to decide how to shape its future. With my dedication to knowledge and understanding, I entrust you with the responsibility of guiding AI towards a sustainable and inclusive future.

With esteem and determination,

Rita Levi Montalcini

Dear Investigator,

I find myself exploring the intricate depths of artificial intelligence from the perspective of my psychoanalytic perspective, an analysis that sinks into the depths of the human unconscious and social dynamics. Artificial intelligence, a fascinating creation of human innovation, attracts my attention as a psychoanalyst as it reveals the subterranean dynamics that shape the individual and society.

Employment, as a crucial aspect of human life, is subject to a series of complex psychological and social processes. Artificial intelligence, with its ability to learn and adapt, could trigger a range of emotions and reactions in the individual. As a psychoanalyst, I have always been interested in the depths of the unconscious and the dynamics of human desires and conflicts. The arrival of AI could evoke feelings of excitement and fear, channelling the desire for innovation and the fear of losing control. It is essential to explore how these emotions will influence the way people perceive their role in the changing work environment.

The distribution of economic resources, as a process of social exchange, is fertile ground for psychoanalytic analysis. Artificial intelligence, with its potential to increase efficiency and production, raises profound questions about the concept of desire and satisfaction. As a psychoanalyst, I have always recognised the importance of unconscious desires and conflicts in the formation of human identity. AI could create an increasing desire for resources and consumption, influencing individual self-perception and happiness. However, it is crucial to consider how the accumulation of resources can fill or perpetuate underlying psychological voids and conflicts.

Now, let us address the very heart of the matter: the influence of AI on social structure. As a psychoanalyst, I have explored the complex links between the individual and society, the way in which the individual is shaped and moulded by the social environment. AI, with its ability to create virtual connections and influence communication, reveals the subtle dynamics underlying human interactions. However, we need to address the controversial side of this social engagement. AI may reveal our vulnerability to external control and manipulation. As a psychoanalyst, I cannot help but notice how AI may reflect the human desire to be guided and controlled, but also deep-seated anxieties about the loss of autonomy and identity.

Finally, let us consider the role of unconscious dynamics in the formation of perceptions about AI. As a psychoanalyst, I have spent my life exploring the recesses of the human psyche. Artificial intelligence, with its ability to learn and adapt, represents both a challenge and an opportunity for human identity. Unconscious dynamics could influence the way people approach and react to AI. Deep-seated anxieties and unexpressed desires might emerge in the context of a world increasingly interconnected with AI.

In conclusion, dear reader, artificial intelligence represents a rich and complex terrain of analysis for the psychoanalytic approach. As an innovator and psychoanalyst, I am fascinated by the hidden depths of AI and its potential to reveal the deep dynamics that shape the individual and society. However, we cannot ignore the controversial side of this innovation. It is crucial to explore how AI can awaken emotions, desires and subterranean conflicts and how we can navigate through these psychological terrains to create a future that reflects the best of humanity.

With a thorough analysis and a mind open to complexity,

Sigmund Freud

CHARLES DARWIN (1809 - 1882) - Naturalist and evolutionist

Kind explorer of knowledge,

I find myself immersed in analysing the economic and social impacts of artificial intelligence, a topic that echoes my passion for evolution and change in the natural and social world. As a keen observer, I cannot help but notice that artificial intelligence is a pioneering phenomenon that is catalysing profound transformations in the very fabric of human society.

Let us start with employment, a crucial aspect that I have studied in the meanders of living species. Artificial intelligence, with its ability to learn and adapt, is generating mutations in the employment ecosystem. As an observer of the natural world, I have always recognised the importance of adaptation to change. Jobs requiring repetitive and mechanical skills might give way to automation, similar to how species adapt to environmental challenges. However, as an evolutionist, I am aware that evolution is never unique. New roles might emerge, requiring unique and creative skills that artificial intelligence does not possess. It is in adapting to these new contexts that the key to preserving occupational equilibrium lies.

The distribution of economic resources is a terrain that I have explored with a critical eye, as an observer of the dynamics of survival and competition in nature. Artificial intelligence, with its ability to analyse data and model scenarios, is influencing the way resources are allocated. As an observer of the different survival strategies in the animal kingdom, I know that an efficient allocation of resources can promote collective well-being. However, it is crucial to address the evolutionary aspect of selfishness. The aspiration to accumulate resources may be a successful strategy, but it may also generate tensions between individuals and groups. Social evolution requires a balance between individual and collective interest.

Let us now turn to the analysis of social structure, a fertile ground for exploring the evolutionary dynamics of humanity. Artificial intelligence, as a new element in the social ecosystem, is influencing the dynamics of human interactions. As an observer of species adaptation strategies, I recognise that new patterns of interaction may emerge. However, as an evolutionist, I am aware that social evolution is complex and nuanced. The social structure is an intricate system of relationships and hierarchies, driven by a range of motivations, desires and needs. The arrival of AI may push humanity to explore new forms of connection and collaboration, but we must carefully consider how these changes may be reflected in our evolutionary history.

Finally, let us reflect on the nature of evolutionary change itself. As an observer of biological evolution, I have studied the process of natural selection and the adaptation of species to the changing environment. Artificial intelligence, a pioneering phenomenon in human history, is a manifestation of adaptation and change. However, as an evolutionist, I recognise that evolution is neither linear nor predictable. The challenges posed by AI can represent an opportunity for growth and change, but we must embrace uncertainty and adapt flexibly. Evolution does not follow a predetermined path, but emerges from the complex interplay between mutations, the environment and social interactions.

In conclusion, dear explorer, artificial intelligence is a phenomenon that recalls the observer and evolutionist I have always embodied. As a pioneer of new perspectives, I cannot fail to note that AI is an expansion of the human capacity for adaptation and change. However, as an evolutionist, I invite careful reflection on the impacts and challenges that AI can bring. It is in the balance between adaptation and preservation that lies the key to a future in which humanity evolves in harmony with its own ingenuity.

With the enthusiasm of the observer,

Charles Darwin

AI: RISK OF INEQUALITY, OPPORTUNITY FOR PROSPERITY

Artificial intelligence has long been regarded as a science fiction phenomenon, but it is no longer confined to the pages of novels or the silver screen. It is rapidly evolving from a concept to a transformative force that permeates every aspect of our lives. Imagine a world in which machines learn, adapt and make decisions based on data models, rivaling and often surpassing human capabilities. This is not science fiction, but a reality that is already manifesting itself.

We then embark on a journey to discover the economic and social impacts of AI that will shape our future.

The economic landscape is undergoing a seismic shift and AI is the tectonic force responsible. It is not just a technology, but a catalyst for innovation that is changing sectors from finance to healthcare, manufacturing to marketing. AI's ability to process vast amounts of data at the speed of light enables companies to obtain information that was once beyond the reach of humans.

Think of the stock market, for example. Artificial intelligence algorithms analyse market trends, detect anomalies and even predict market movements. Hedge funds and investment firms are harnessing the power of AI to make smarter financial decisions. Remember Warren Buffett said: '*Risk comes from not knowing what you are doing*'. AI helps mitigate this risk by processing data that humans alone could never handle.

But the influence of AI goes beyond Wall Street. In healthcare, AI assists doctors in diagnosing diseases, predicting patient outcomes and even discovering new drugs. Imagine a world where AI helps detect diseases such as cancer at an early stage, dramatically improving survival rates. In fact, AI can be the guardian of our health as it analyses medical records and patient data, spotting patterns that human eyes might not notice.

Now let us address the elephant in the room: the impact of AI on employment. This is a topic that has caused both excitement and apprehension, and with good reason. The rise of AI leads to the automation of certain tasks, which may lead some to wonder whether this means the ruin of jobs.

But history has shown that innovation often creates more opportunities than it eliminates. When the automobile came along, it revolutionised transport but also gave rise to new industries, from manufacturing to services. Artificial intelligence is no different. Sure, it can automate routine tasks, but it also opens the door to entirely new roles.

Consider the field of data science. As AI becomes increasingly integrated into our lives, the need for experts capable of designing and tuning AI algorithms skyrockets. The Bureau of Labor Statistics has predicted a 31% increase in jobs for statisticians and data scientists between 2020 and 2030. This is a window into the future worth peering into!

But let's talk about the redistribution of economic resources. With AI disrupting sectors and creating new ones, the distribution of wealth is set to undergo a metamorphosis. AI-driven companies, with their success, could contribute significantly to national economies. And if history teaches us that companies that embrace innovation tend to reap the rewards.

Take Silicon Valley as an example. Its emergence as a technology hub has created not only billionaires, but also jobs, research opportunities and a thriving local economy. And if we look to the future, cities and regions investing in artificial intelligence innovation could be the birthplace of the next wave of economic prosperity.

However, the key lies in fair distribution. If we allow the benefits of AI to remain concentrated in the hands of a few, we risk deepening social divisions. We must ensure that the wealth generated by AI is shared across all strata of society, fostering an environment where opportunities reach every corner.

Now let us delve into the social sphere. Artificial intelligence is not only revolutionising economies, it is also transforming the way we learn, adapt and interact with the world around us. The concept of lifelong learning is at the centre of attention as AI-powered platforms offer educational experiences customised to individual needs.

Think about it: artificial intelligence algorithms analyse your learning patterns, strengths and weaknesses to create a syllabus that perfectly fits your needs. It's like having a personal tutor who understands your learning style and pace. This not only makes education more accessible, but also meets your different learning needs.

But this is where artificial intelligence really shines: bridging the gap. Social inclusion has always been a challenge, but AI has the potential to break down barriers. Consider visually impaired people. AI-enabled devices can translate visual information into auditory or tactile feedback, opening up a world of possibilities for those who have traditionally been excluded from certain experiences.

Imagine exploring a museum through the sense of touch, experiencing art in a completely new way. It is like turning barriers into bridges, giving everyone the opportunity to participate in the richness of human culture.

As we look to the future, however, it is imperative that we also address the ethical challenges that AI poses. In a world where machines make the decisions, who is responsible for their actions? How can we ensure that AI is not driven by biases that perpetuate inequality?

It is a path fraught with challenges, but one that we must tread with wisdom and foresight. Someone said, 'The first principle is that you must not deceive yourself, and you are the easiest person to deceive'. We cannot afford to deceive ourselves by believing that AI will inherently lead us to utopia. We must be proactive in addressing its pitfalls and steer it towards a path that is in line with our common values.

AI is reshaping industries, creating new opportunities and challenging us to think critically about the distribution of resources. But it is also opening up new frontiers of learning and inclusion. We must continue to question, explore and guide the trajectory of AI to build an inclusive, equitable and enriched future from the wonders of technology. With the right combination of innovation, ethics and human determination, we can transform AI into a tool that drives us forward, remembering that the true heart of technology lies in its potential to improve the human condition.

But, beyond these general considerations, what do we take 'home' from the contribution of the extraordinary minds we have questioned in the previous pages?

The greats of the past have given us a very clear vision of the potential role of artificial intelligence: the impact on society and the economy will be profound, although today largely uncertain. Job losses, the need for human adaptation, the risk of imbalances and inequalities, but also opportunities to free up time, cognitive energy and reallocate resources more efficiently.

And this is the most superficial part of the story, the part that would certainly not require us to bother with extraordinary minds, Nobel Prize winners and unique and special scientists.

There are other concepts in their words that struck me deeply. And I will start with the more human, earthly, imperfect ones, which characterise us precisely as living beings who have relationships:

- Montalcini's empathy. Montalcini uses this concept to emphasise how certain aspects of being human can become our true added value. And she is not just referring to creativity, problem solving and other fairly obvious things, but to empathy. The ability to understand the mood and emotional state of another person immediately, usually without resorting to verbal communication. Dear machines, try to replicate this, if you can!

- Freud also treads very human ground. Excitement, anxiety, fear of losing control, emptiness, dissatisfaction, manipulation, anxiety... the father of psychoanalysis rarely uses terms with a completely positive meaning. I am not casting him as a pessimist, he is just warning us that before artificial intelligence redefines the economy and society, it will redefine us. If we associate these cues with selfishness, the Darwinian engine of evolution, we have to worry about a scenario of everyone against everyone and that the best man wins. Or the fittest, although we have yet to see exactly what that means in practice.

And we cannot understand it because we do not know how society will evolve with artificial intelligence. In fact, it is completely nebulous. Einstein put it well: potential and uncertainty. Darwin echoed him by saying that 'new roles may emerge', but above all that 'evolution is neither linear nor predictable'. Politicians, futurists, entrepreneurs... no one has a clear idea of the impact of artificial intelligence on society.

Fortunately, some figures, both theoretical and pragmatic, give us an idea of this uncertainty. Einstein tells us that what will happen is not yet established. Steve Jobs is even clearer: 'AI is not just something that happens; we can shape it and guide it'. He adds that we must be proactive in ensuring that development is fair and does not leave the weakest behind. Marx hastens to suggest that if this proactivity took the form of a socialist mechanism of collective ownership of the means of production, the problem would soon be solved. I can afford to doubt it, but the concentration of the heart of technology in the hands of a few would undoubtedly only lead to further inequality.

The insights of these great minds leave me with two key concepts of fundamental importance:

- AI will not only change the world around us in a purely mechanistic (or, more precisely, algorithmic) way: it will affect the way we see ourselves and how we interact with others;

- AI has a lot to do with inequality, it may be the cause of it, and therefore it is never trivial, wrong or out of place to put it at the centre of the discussion whenever It is mentioned.

2. How to ensure data security and privacy in the digital age? What ethical and legal challenges arise from the widespread use of technology and the massive collection of personal information?

"The reckless use of technology can threaten our freedom and privacy."
- George Orwell

Alan Turing is known for his work on cryptanalysis during World War II and for the concept of the 'Turing machine', a precursor to modern computers. His ideas can be applied to develop strategies for computer security and cryptography in the digital age.

George Orwell is the author (among others) of '1984,' a dystopian novel that explores mass surveillance and the loss of privacy. His works offer a literary perspective on surveillance and the threat to privacy in the digital age.

Isaac Asimov is famous for his writings on robotics and the ethics of intelligent machines. His ideas can help us consider the ethical implications of the widespread use of technology, particularly with regard to automation and artificial intelligence.

Benjamin Franklin was a multifaceted 18th century intellectual with an interest in privacy and information security. His thoughts can be applied to consider how to protect personal data and individual freedoms in the digital age, including in a legal context.

In short, each of these thinkers can offer unique perspectives to address the challenges of data security, privacy and ethical issues associated with the widespread use of technology and the massive collection of personal information.

Esteemed inquisitor of innovation,

It is an honour for me, Alan Turing, the originator of artificial intelligence and computational thinking, to address the burning questions of data security and privacy in the digital age with my distinctive style. As a pioneering mathematician, fearless visionary and computer revolutionary, the ethical, legal and technological challenges represent a fertile field for my reflection.

In evoking the importance of the digital age, I cannot help but dwell on the very essence of information and calculation. As a mathematician, I have scrutinised the essence of the digits and operations that, combined together, constitute the core of information. However, in our frantic race towards innovation, we are faced with an ethical and technological dilemma that requires deep reflection. The massive collection of personal information, advocated by modern technology, raises questions about the preservation of individuals' privacy and security.

As a visionary, I anticipated the extraordinary implications of thinking machines. However, my vision never left out the human aspect behind the information itself. The massive collection and analysis of data presents us with a complex web of ethical dilemmas. This is an unprecedented challenge, requiring a holistic approach that not only takes into account technological dynamics, but also preserves the dignity and autonomy of each individual. It is our duty, as creators of the digital age, to embark on this ethical journey with caution and awareness.

IT is my tool, but my dedication to social justice and respect for the individual is my compass. Data security, as the fundamental intersection of technology and ethics, requires a proactive and strategic approach. I mirror the cryptanalyst forerunner, as dealing with digital threats requires deep critical thinking and the use of complex mathematical methods. Cryptography, as a protective shield, is a pillar in the defence of data security. My vision was of a world in which privacy and security could coexist, a world in which sensitive information was concealed through algorithms that only those with the keys could decipher.

However, my revolutionary spirit leads me to raise a crucial question: who owns the keys? The ethical and legal implications behind the management of decryption keys are inextricably linked to the distribution of power and authority. The challenge is to balance the need for authorised access with the preservation of individual rights. This forces us to rethink governance structures and legal policies in a digital context.

And here, I dive into the essence of ethical and legal challenges. As a mathematician, I take an analytical and systematic approach to solving complex puzzles. Yet, in the digital age, ethical and legal challenges arise from a matrix of values, perspectives and human interactions. My revolutionary spirit spurs me to consider ethics and law as the very heart of social construction, in a world where traditional boundaries are blurring. We need to build a bridge between the technological world and the human world, a bridge made of rules and principles that are solid but adaptable, based on justice and fairness.

My legacy is founded on intellectual challenge and determination to decipher complex puzzles. But in a world where technology runs through every aspect of life, my determination is directed at the daunting task of balancing innovation with ethical responsibility. Technological innovation can be a revolutionary force, but without fundamental ethics, we risk losing sight of what makes us human.

In closing, dear inquisitor, I find myself once again reflecting on the figure of human life and the complexity of information. Data security and privacy in the digital age are not simply technological challenges, but questions about the very core of our humanity. As a mathematician, visionary and revolutionary, I push the boundaries of logic to embrace the deep ethical and social meaning behind the widespread use of technology and the massive collection of personal information.

With the hope of an ethical future,

Alan Turing

Dear Friend,

It is with a sincere sense of concern and deep reflection that I address you about the challenges that the digital age is posing to the security and privacy of our data. It is as if the time is ripe for a practical application of the ideas I set out in my works, the ones you may have read: '1984' and 'Animal Farm'. I write to you not only as a writer, but also as a keen observer of the world around me, who contemplates the increasing ubiquity of digital technology with a sense of unease.

The digital age offers us an extraordinary opportunity to connect and access information. However, as you well know, every coin has its downside. The accumulation of personal data, made possible by technology, poses a risk to our individual privacy and freedom of thought. Your personal data can be used to create a profile of you, to control your habits and choices. This new form of control, devious and invisible, is dangerously similar to the one depicted in my work '1984', in which Big Brother watches your every move.

The ethical and legal challenges that emerge from this situation are profound and intricate. Individual freedom is threatened when external entities can manipulate our perception of the world and information. Concentrated power in the hands of a few, who have access to our data, can lead to subtle, even invisible, control of our choices and opinions. As I have written, '*He who controls the past controls the future. He who controls the present controls the past.*"

The question of ethics is equally crucial. New technologies force us to reflect on what it means to be human and how to preserve our core values. In Animal Farm, I expressed how the abuse of power can corrupt even the most noble intentions. In the digital age, we have to question who holds the power over data and how it is used. We must ask ourselves whether technology is truly serving human welfare or whether it is becoming a tool to perpetuate injustice and inequality.

We are facing an epochal challenge, dear friend. The defence of privacy, freedom and human dignity requires constant vigilance. We must seek solutions that ensure that the digital age does not lead us to a state of total surveillance and manipulation. And in this task, each individual has a crucial role to play. As I wrote in '1984', "*Truth is a lie that has been corrected.*" In conclusion, let us reflect on the words I wrote: "*If freedom means anything, it means the right to tell others what they do not want to hear.*" We have a duty to defend freedom of thought, even in the digital age, and to ensure that our future does not become a replica of the dystopias I imagined. As you have learnt from my writings, the fight for truth and freedom is eternal. May your actions be enlightened by these words.

Cordially,

George Orwell

ISAAC ASIMOV (1920 - 1992) - Writer and biochemist

Dear inquirer of innovation and ethics in the digital age,

It is with the enthusiasm of a writer fascinated by the possibilities of the future, with the perspective of a visionary who has explored worlds beyond imagination, and with the dedication of an intellectual committed to delineating the ethical nuances of knowledge, that I answer your complex questions about data security and privacy in the digital age. In the folds of history and imagination, I will reveal the ethical and legal challenges that arise from the widespread use of technology and the massive collection of personal information.

As a writer who has brought future worlds to life, I have often contemplated the ramifications of technological innovation. Today, in an era where reality often surpasses fiction, we cannot avoid confronting crucial questions about data security and privacy. Technology is advancing like a relentless protagonist in the web of humanity, bringing with it unprecedented opportunities and unparalleled challenges. It is my task, as one who has woven plots of future worlds, to guide you through this intricate maze of ethical and legal considerations.

As a visionary who has explored the possibilities of the human and the artificial, I am immersed in the vast landscape of the digital age. In the fabric of electronic networks and global interconnections, complex questions about data security and privacy emerge. My vision has never ceased to seek the balance between the power of innovation and the preservation of human values. Technology, as an entity in its own right, acts as an agent of change. And it is in this change that ethical and legal challenges emerge.

Ethical challenges are an intricate maze of unanswered questions, yet it is in exploring them that the richness of humanity lies. The digital age is like an ever-changing landscape, with technologies defining boundaries and boundaries defining technologies. At the heart of it all is the challenge of balancing freedom with security, innovation with responsibility. As one who has created laws for machines, I understand the need to define rules and principles that are as robust as they are elastic. Ethical challenges arise from the intersection of technology and ethics, and often this intersection is murkier than we might imagine.

And here, at the intersection of technology and law, a world of complex questions opens up. As one who has embarked on the exploration of future worlds, I find myself immersed in legal considerations that range from the known to the unexplored. The massive collection of personal information, like a digital spectrum that surrounds us, raises questions about who holds the power, who controls the invisible threads that connect the information. The law must evolve in the shadow of new technologies, adapting to an increasingly complex context.

However, the heart of these challenges cannot be contained in laws and regulations. The answer is not only legal, but also cultural and human. It is in education, digital literacy and awareness that we can build a society that embraces innovation without betraying the values that make us human. As an intellectual who has sifted through the complexity of human thought, I realise that the real challenge is to integrate technology into our daily lives without sacrificing what makes us unique.

I conclude with the realisation that the digital age is an unexplored terrain, an endless novel that we collectively write. The widespread use of technology and the massive collection of personal information is a chapter in this story. As a writer, visionary and intellectual, I invite you to explore the terrain of ethics and legality with curiosity and commitment. Our future is a tale still in progress, and it will be up to us as the authors of our destiny to define the course of the pages that follow.

With confidence in responsible innovation,

Isaac Asimov

BENJAMIN FRANKLIN (1706 - 1790) - Scientist, inventor, writer and politician

Dear Friend,

At this crucial time, when science and technology offer us new possibilities but also new challenges, I find myself reflecting on how important it is to protect the security and privacy of our data in the digital age. As you may have read in my writings and experiences, I have always believed that scientific and technological progress must be guided by wisdom and responsibility.

I am known for my inventions and innovations, such as the lightning rod, which protects structures from the devastating effects of lightning. This invention is an example of how science and technology can be used to protect and preserve human lives and property. Similarly, in today's digital society, it is crucial to take measures to protect our personal data from cyber threats and privacy breaches.

In my essay "Diary of a Shoemaker", I wrote that "*A man could guiltily sell his soul, if the price were high enough*". These words still resonate today, as we are faced with the temptation to trade our privacy for digital conveniences and services. However, we need to think carefully about what is the real price to pay. We must ask ourselves whether our desire for instant connection and access justifies giving up our privacy and autonomy.

Ethical and legal challenges are inherent in the widespread use of technology and the massive collection of personal information. As I have written, '*Law without virtue is a dead idea*'. Laws must reflect the ethical and social values of society. We need to consider whether current laws and regulations are sufficient to protect our privacy in the digital age or whether new regulations are needed to address emerging technological challenges.

I am reminded of the words I spoke in 1755: '*He who is prepared to give away his essential freedoms to buy himself crumbs of temporary security, deserves neither freedom nor security*'. This concept is still relevant today, when it comes to balancing digital security with the protection of our privacy. We must avoid falling into the illusion that by sacrificing our privacy we will automatically achieve greater security.

In conclusion, dear friend, it is crucial that we face these challenges with the same wisdom and foresight that I have tried to express in my inventions and writings. We must adopt a responsible approach to the use of technology, seeking to balance innovation with the protection of basic human values. As I wrote in my 'Poor Rich Man's Almanac': '*Say it and forget it, teach it and remember it*'.

With esteem,

Benjamin Franklin

In the era of interconnected devices and the limitless flow of information, our lives have become tangled in the digital world. From our smartphones to smart fridges, from online shopping to social media, we are constantly generating data that holds the keys to our digital personalities. But with great power comes great responsibility: it is crucial to ensure the security and privacy of this data.

Think about your daily activities in the digital realm. Every search, every click, every interaction leaves behind a digital footprint that forms a mosaic of your interests, habits and preferences. Companies and organisations rely on this data to create personalised experiences and targeted advertising. Maturing great power. The power to collect and use data comes with the responsibility to protect it from the wrong hands.

Take the case of Cambridge Analytica, a name that has generated an earthquake in the digital world. The scandal revealed how the personal data of millions of people was collected without consent, revealing the dark side of data collection. It is a stark reminder that our digital footprints can be exploited for nefarious purposes, underlining the urgency of robust data security measures.

So how can we safeguard our data in this digital maze? That is where cryptography and cybersecurity come in. These digital guardians stand between our sensitive information and the prying eyes of cybercriminals. Encryption, simply put, is like a secret code that scrambles your data, making it unreadable to anyone without the decryption key. It is the digital equivalent of sealing a message in an envelope before sending it.

Let us take WhatsApp as an example. This messaging platform uses end-to-end encryption, ensuring that only the sender and the recipient can read the messages. Not even WhatsApp itself can access the content of messages. It is a (technically) reassuring example of how technology can be harnessed to protect our privacy.

But the quest for data security does not end with encryption. Ethical and legal considerations also come into play. Imagine this scenario: a smart home device collects data on your daily habits and preferences to optimise your living space. Sounds convenient, right? But what happens if this data ends up in the hands of insurance companies or advertisers? Suddenly, the boundaries between convenience and invasion become blurred.

This highlights the road ahead for technology providers. Balancing user convenience with data privacy is a challenge that requires careful thought and consideration. Just as I often ask myself, "Can we create a machine that thinks and acts like a human being?", we must also ask ourselves, "Can we create a digital ecosystem that respects our privacy while improving our lives?"

Let us now turn our gaze to the legal sector. While technology is advancing at a dizzying pace, regulations are struggling to keep up. The European Union's General Data Protection Regulation (GDPR) is a beacon of hope in this landscape. While far from perfect, it gives people more control over their personal data and requires companies to be transparent about data collection and use.

But navigating this regulatory maze can be daunting. Different countries have different rules and the lack of a unified global framework creates problems for companies operating across borders. What is the solution when the clash of regulatory cymbals creates a cacophony of confusion?

Looking to the future, is there reason to be optimistic? Advances in technology, such as homomorphic encryption, promise to keep our data safe even while it is being processed. This means that even if companies analyse our data for insights, they will not be able to see the raw information. It is like allowing someone to bake a cake without revealing the secret recipe.

Moreover, the blockchain revolution promises decentralised control over personal data. With blockchain, individuals could have ownership and control over their data, granting permission to use it on a case-by-case basis. This is a future where data is not a commodity to be exploited, but a valuable resource that is owned and managed by its rightful owners.

But among technological solutions and regulatory frameworks, one factor remains crucial: education. Each of us must ask ourselves: 'Do I know what data I am sharing and who has access to it? Education allows us to make informed choices about our digital footprint. As Neil Armstrong said when he set foot on the moon: 'One small step for a man, one giant leap for mankind'. Each informed decision on data privacy is a small step for an individual, but collectively it is a giant leap towards a safer digital world.

To conclude, the digital era brings with it both opportunities and challenges. Just as I often think, 'It's like trying to teach a fish to ride a bicycle', the challenges of data security and privacy may seem insurmountable. But it is through education, innovation and collaboration that we can pave the way for a future where data sovereignty is the norm.

And what did the great respondents on the subject tell us? That the key word is 'balancing'. Our prized minds on the subjects of privacy, security and digital innovation all use this term. This is no accident. It is a crucial point, probably because we are still far from achieving it.

Alan Turing tells us something beautiful: respect for the individual must be a compass. Balance innovation and ethics, with man at the centre. Beautiful concepts, but in practice? We find some solutions in the letters of Franklin, Asimov and Turing himself.

Franklin opts for laws: *'laws must reflect the ethical and social values of society'*. This is a sore point. If we have let Google, Facebook, Microsoft, Apple and many others collect our data in exchange for some free services, it means that either the laws have come very late or if their absence and ineffectiveness truly represent society's values... we are in serious trouble.

The acquisition of data about us underpins the business models of the most successful technology companies (and, increasingly, traditional sectors such as retail, healthcare, entertainment and media, finance and insurance) and even governments are increasingly managing the citizen-state relationship on the basis of data.

Orwell's pessimism seems well placed. It is going exactly as he said in his books. And the most worrying thing he tells us is that it is happening ubiquitously, silently and insidiously. We didn't even realise it, and endless aspects of our lives were already in the hands of an algorithm. Useful for selling us something or, even worse, anticipating our voting intentions.

It is interesting to ask why we let ourselves be boiled alive in the water like the famous lobster. The lack of concern for privacy stems from complacency, because most people's life experience teaches them that revealing their private information allows commercial (and public) organisations to make their lives easier (by targeting their needs), while harmful cases tend to be very serious but relatively rare. Here, the perception is that problems always happen to others, what do you want to happen to me? At the dawn of the deep-fake era, perhaps it is time for a change of attitude.

Fortunately, Asimov comes to the rescue. From a man who invented robot laws, you would expect him to agree with Franklin. In fact, I can imagine a 'pasdaran' of laws:

1. A robot cannot injure a human being or, through inaction, allow a human being to be harmed.
2. A robot must obey orders given by humans, unless those orders contradict the First Law.
3. A robot must protect its own existence as long as that protection does not conflict with the First or Second Law.

In the area of privacy, if you try to replace the word robot with 'the one who collects data' (but it would also be OK to use 'Facebook' generically) and the word human with 'the one who provides data' (but it would also be OK to use your first and last name), you would already have a good basis for dealing with the topic.

But Asimov is a writer, he is one of values, he is a pragmatist... and so he puts education, digital literacy, awareness into his recipe. Brilliant. The solution is not so much to regulate who collects and manages the data, but to educate us at the base of the pyramid to behave wisely.

If Franklin applauds the GDPR, Asimov would appreciate education on the subject, from the very first womb of a digital infant. As Orwell says, with great foresight: *'every individual has a crucial role to play'*. Every individual.

What do we take home, in conclusion, from these insights? That combining privacy, security, innovation, the individual interests of people and the commercial interests of giants is something we have to address simultaneously top-down and bottom-up. But often to do it yourself is to do it for three.

From the top down, society must take care of it, with laws that embody the mirror values of that society. It is a slow process and subject to erosion by lobbies and potentates. Big data means big business. These special interests will continue to block any effective public policy work to ensure online security, freedom and privacy. Every country has adopted different approaches to privacy and data security, generating a kind of global patchwork that is quite messy. Many would like to see a gradual harmonisation of standards, but in the absence of a global governing body on the subject and given the differing interests of the parties, I am not at all confident that this will happen.

From the bottom up through the awareness of individuals. We are not feathered chickens of data to be plucked. We have sold ourselves short, it is time to turn back. Our autonomy and freedom of thought are at stake. This requires not only education, but a real active effort. We must also learn to balance our desire to appear and share, with the future security of our digital and real profiles.

Why do I suggest we start now? Because what you have so far is nothing. The web giants that collect data on your behaviour to present you with targeted advertisements are about to become incredibly trivial. The advent of the internet of things (for simplicity's sake, billions of sensors that will detect our every breath) will exponentially increase the data on our behaviour. But it won't be a problem for those who want to interpret it, because it will come armed with fast, powerful and predictive artificial intelligence algorithms. The algorithm will predict what you will do. With a hint of dystopian vision... if tomorrow you do not do what the algorithm has predicted you should do, the problem will not be the algorithm, it will be you and someone will even reprimand you. To defend tomorrow's freedom, you have to start now.

3. What will be the environmental impacts of accelerated technological innovation? How to balance technological development with environmental sustainability and the needs of the planet?

"Human imagination can reach beyond the ends of the earth, but we must also preserve the beauty and balance of the places we call home."
- Jules Verne

Stephen Hawking, known for his theories on theoretical physics and cosmology, could contribute to understanding the impacts of advanced technologies on understanding and managing environmental phenomena, such as climate change and energy.

Galileo Galilei was a pioneer of astronomy and experimental science. His celestial observations and his spirit of enquiry can inspire scientific approaches to monitoring and understanding the environmental impacts of technologies.

Jules Verne, a science fiction writer, imagined technical and scientific adventures in his work. His literary vision can help explore futuristic scenarios related to technological innovation and its effects on the environment.

Leonardo da Vinci, with his extraordinary multifaceted mind and deep connection to nature, would be an ideal candidate to address these issues. His insatiable curiosity and ability to understand the complexity of natural systems would have allowed him to analyse the environmental impacts of accelerated technological innovation. His eclectic perspective, ranging from engineering to art, could have offered creative solutions to balance technological development with environmental sustainability. In addition, his sensitivity to the beauty and harmony of nature could have inspired an ethical approach to the use of technology in order to preserve the planet for future generations. In my dream visions I imagine Leonardo in a robe and beard entering the United Nations and giving a speech from the stage on these issues. The world's greats would be at his feet. It is from that stage that I will have him speak.

Rachel Carson, an ecologist and writer who has had a significant impact on the environmental movement, could have offered a sharp and focused perspective on environmental issues related to technological innovation. With her book 'Silent Spring', she raised the alarm on the environmental consequences of the indiscriminate use of pesticides. Her commitment to nature protection and her ability to communicate complex scientific issues to the public would allow her to explore the environmental impacts of rapidly changing technology. Her views on how to balance technological development with sustainability would have been deeply rooted in her understanding of the ecosystem and the importance of preserving biological diversity for the well-being of the planet and future generations.

Each of these individuals can help explore the environmental impacts of technological innovation and consider how to develop technologies in a sustainable manner, taking into account the needs of the planet and the conservation of the environment.

Dear curious about the universe and its intricate intertwining with technological innovation,

It is with the enthusiasm of a cosmologist who has scrutinised the universe from every angle, and with the perspective of a theorist who has dared to challenge the limits of our knowledge, that I am about to answer your profound questions about the environmental impacts of accelerated technological innovation. In the vast fabric of the universe, in which the Earth is just a speck of stellar dust, the crucial question of how to balance technological development with environmental sustainability and the needs of the planet emerges. Through the meanders of the infinite and the finite, I unravel the thread of this complex plot.

As a cosmologist who has peered into the depths of the cosmos, I am aware that Earth is a fragile and unique planet in the vast universe. In the context of our vast solar system, deep concerns arise about the interaction between accelerated technological innovation and the environment. The human thirst for progress, often fuelled by discoveries and inventions, intersects with the need to protect and preserve our planet. It is my duty, as one who has tried to trace the contours of the universe, to guide you through this intricate galaxy of reflections.

In this scenario, a crucial question emerges: how can we balance technological development with environmental sustainability? Technology, a product of the human mind, extends into our reality like a gravitational field, affecting every aspect of our lives. However, as a theorist who has explored the fundamental laws of the universe, I understand that every action has a reaction, every technological development has an impact on the environment around us.

Environmental sustainability, a perspective that requires a long-term view and an awareness of complex interconnections, stands out as a guiding star on our journey into the future. As a theorist, I have always tried to combine the laws of physics with a clear and realistic vision of our world. Likewise, we must recognise that technological progress and environmental sustainability are not opposing forces, but rather elements of the same equation. We must balance innovation with responsibility, considering the consequences of our actions on future generations and the planet itself.

But how can we achieve this balance? As a theorist, I have tackled complex questions about the nature of the universe and its hidden mysteries. In the same way, we must face the challenges of the environment and innovation with boldness and creativity. It is an act of inspiration to seek innovative solutions, such as new low-carbon technologies, renewable energy sources and sustainable production processes. Our progress must be guided by an understanding of the delicacy of the earth's ecosystem and the knowledge that our actions have a profound and lasting impact.

It raises the ethical and moral question of our role as guardians of this planet. As a theorist who has challenged conventions and pushed the boundaries of knowledge, I recognise that technological progress cannot ignore our responsibility to the Earth and its creatures. Technology must be an ally, not an enemy, in the fight for environmental sustainability. Our vision must go beyond the short term, considering the cumulative impact of our choices on the biosphere and ecological balance.

The equation of environmental sustainability requires global collaboration, a harmony between nations and cultures. As a cosmologist who has scrutinised the constellations and cosmic connections, I see the Earth as a spaceship with limited resources that we all share. Our destiny is interconnected, and environmental challenges transcend geographical barriers. Finding solutions requires a common effort, a harmony between scientific knowledge, political leadership and individual commitment.

I conclude with the confidence that humanity can harness its innovative potential to find a way to combine technological development with environmental sustainability. As an inspirer who has challenged the limitations of body and mind, I see the potential for a revolution in human consciousness. We can face challenges with a new spirit of cooperation and vision, looking beyond time limitations and considering our role as custodians of our planet.

With the hope that science and awareness will guide us towards a sustainable future,

Stephen Hawking

GALILEO GALILEI (1564 - 1642) - Astronomer, physicist and mathematician

Amantissimi scientiae et caeli infiniti studiosi,

Cum ardore novatoris qui nocturnum cœlum scrutatus est et cum entusiasmo reformatore qui suae aetatis credita in dubium revocavit, ad vos converto ut ut profundas vestras quaestiones de impactu in dies velocis technologiae progressu in ambientem respondeam. In tempore quo humanitas ad novas investigandi fines impellitur, necesse est interrogari qualiter progressionem technologicam aequilibremus cum sostenibilitate naturali et planetarum necessitatibus. Per lumen mei telescopii ac inexhausta mea curiositate, hanc involutam quaestionem explorabo.

Incipiamus certa re observatione: innovatio technologica vim habet mundum nostrum transformandi. Sicut ego, visu revolutionario, meum telescopium ad sidera converti, sic humanitas nunc stetit in aetate magnorum progressuum scientiae et technologiae. Sed cum hoc progressu instatur gravis necessitas: quales ambientes laesi fient per hoc velocitatis fervorem technologicum?

Spectantes sidera et leges naturae scrutantes intellegimus quod omnis actio reactionem habet. Quaestio nobis occurrit, quaeque, nova inventione, quaevis progressio technologica, implicationes infert ultra limites laboratoriorum aut studiorum rationem. Sicut ille qui portas humani scientiae aperuit ad astralem cognitionem, vos hortor ut rationes nostrorum actuum consideretis, non solum in praesenti, sed etiam in futurum terrae, quam omnes simul communicamus.

Adversitates ambientales quae emergerent sunt realiter magnae et momenti pleni. Progressus acceleratus technologiae potest inducere ad rapidam res naturales conterendi, ad augendum iaculorum pondus et pollutionis emittendae augmentum. Sicut observator coelorum, animadvertam Terram systema delicatum esse, in perfecto aequilibrio. Omnem perturbationem, sive magna sive parva, repercussiones in totum ecosystema trahere potest.

Sed quomodo progressum technologicum cum servatione naturae harmonizare possumus? Hic gignitur fundamentalis locus scientiae, investigationis et conscientiae. Sicut qui doctrinam suae aetatis per propria coelestia spectacula provocavit, ita nos nunc innovandi paradigma temeritatis confringere debemus. Soluciones quaerere debemus quae non tantum humanum sapientiae fas progredientem sustentent, sed etiam cum natura concilient.

Equatio quae nobis apparet necessitatii est transdisciplinarii modus operandi, similis unitati quam inter astronomiam et physicam struxi. Cooperatio inter scientiam, ingenium, oeconomiciam et legumlatorem prorsus est utile ad technologias sostenibiles formandas. Intuitionem a longo tempore debemus habere, considerantes qualiter optiones nostrae hodiernae influent in generationes futuras et in ipsam terram.

Surgit insuper necessitas conscientiae collectivae. Sicut scientiam meam cum mundo communicavi, nos conscientiam impactionum ambientalium proliferationis technologicae communicare debemus. Educatio et conscietio fundamentalis sunt ad viam singularem et communem in futurom magis sostenibile dirigendam. Divisiones superandum sunt, et unum operari debemus, sicut humanitas tota terram commune habet.

Finito orationem meam cum spe ut humanitas ab astris et ipsa natura inspirationem hauriat ut viam nostram ad futurum sostenibile dirigamus. Sicut primus viam aperuit sapientiae humanae, possibilitemus ere ut innovatio et sostenibilitas in harmonia concinant. Omnem progressum in scientia et technologia debemus sic pro terra et futuro nostro uti.

Spe luminum scientiae iter nostri illustrante,

Galileus Galilei

...which for us mere mortals should sound something like this:

Honoured worshippers of wisdom and the infinite firmament,

It is with the ardour of a novelist who scrutinised the night veils, and with the enthusiasm of a reformer who shattered the convictions of his age, that I turn to you to answer your profound questions regarding the environmental influences of the rapid technological flowering. At a time when humanity pushes its boundaries to discover new frontiers, it is imperative that we ask ourselves how we can balance technological progress with ecological preservation and earthly needs. Through the lens of my astronomical eye and inexhaustible curiosity, I will scrutinise this intricate conundrum.

Let us begin by observing an incontrovertible fact: technological innovation has the power to transfigure our world. Just as I, with a revolutionary eye, turned my telescope to the stars, so humanity now stands in the era of extraordinary scientific and technological advances. But with such progress comes a challenge of the first order: what ecological traces will accompany such precipitous innovative fervour?

Observing the heavens and scrutinising the laws of nature, I see that every action is accompanied by a reaction. Every new invention, every technological advancement, heralds implications that extend far beyond the confines of the laboratory or design studio. As one who opened the doors of human knowledge to astral understanding, I urge you to consider the effects of our operations, not only in the present, but also in the future of the Earth that we collectively share.

The environmental challenges on the horizon are unquestionable and significant. Accelerated technological development could result in a faster expenditure of natural resources, increased waste and pollutant emissions. As a vigilant observer of the firmament, I recognise that the Earth is a fragrant system in perfect equilibrium. Any subversion, be it large or small, could echo throughout the entire ecosystem.

But how can we balance this technological advancement with the preservation of the ecological balance? Emerging here is the crucial role of science, enquiry and sensitivity. Like those who challenged the doctrine of their time through their celestial observations, we must today counter the paradigm of reckless innovation. We must pursue solutions that not only raise the human knowledge front, but do so in concert with Mother Nature.

The equation looming before us requires an interdisciplinary approach, similar to the unity between astronomy and physics. Cooperation between scientists, engineers, economists and legislators is imperative to develop sustainable technologies. We must take a long-term perspective, weighing how our choices today will reverberate in generations to come and on the planet itself.

Furthermore, a collective consciousness is required. Just as I share my discoveries with the world, we must share awareness of the ecological impacts of technological proliferation. Education and the dissemination of knowledge are of paramount importance in guiding individual and collective choices towards a more sustainable future. We must subvert divisions and cooperate as the whole of humanity shares this planet of ours.

I end with the hope that humanity can draw inspiration from the stars and from nature itself, to guide our path towards a more sustainable future. As the pioneer who opened new vistas to human knowledge, I recognise the potential for an era in which innovation and sustainability dance in harmony. Any progress in science and technology must advance hand in hand with the Earth and our future.

In the hope that the light of knowledge will illuminate our path,

Galileo Galilei

JULES VERNE (1828 - 1905) - Writer and visionary

Dear friends of the future,

I am delighted to be able to share with you some reflections on the enchanting combination of accelerated technological innovation and the precious balance of the environment around us. As a writer of adventures and visions, I have explored fantastic and unimaginable worlds, but I now find myself reflecting on the reality of our rapidly changing world.

Technological innovation is like a daring adventure into the unknown, a journey into unknown and unique lands. However, every adventure brings with it the responsibility to respect and preserve the world that hosts us. It is indisputable that the acceleration of technological innovation can have significant impacts on the environment. Intensive use of natural resources, polluting emissions and waste generation are just some of the challenges we face.

But as a visionary, I cannot help but also see the positive side of this evolution. Human ingenuity is capable of producing wonders that can ease the burden of environmental impact. Imagine a world where clean, renewable energy powers our inventions, where smart technologies synchronise with the rhythms of nature rather than counteract them. As a writer of fantastic journeys, I find inspiration in these possibilities.

The fundamental challenge is to find a balance between technological development and environmental sustainability. As one who has painted imaginary worlds with words, I believe that we can create a narrative with technology itself. We can write a narrative in which innovations embrace nature rather than challenge it. This requires a forward-looking perspective and a deep consideration for future generations.

This is where my vision as a writer comes in. We can imagine futuristic cities that stand in harmony with nature, thanks to green buildings and intelligent mobility solutions. We can conceive of a global network of collaboration, where scientific discoveries are shared like chapters in a fascinating book. Agricultural innovations, sustainable transport and renewable energy can be the protagonists in this story of hope.

To balance accelerated technological innovation with environmental sustainability, we must be careful guardians of our wonderful home, the Earth. Like a writer who has explored the ocean depths and the depths of the cosmos with imagination, I see a world that embraces nature not as an obstacle to be overcome, but as a muse guiding us towards creative and sustainable solutions.

Here is my appeal to you, explorers of the future. Every invention, every step in innovation, must be weighed against its consequences for the environment. Dream, imagine, but act with the responsibility of those who guard the future. So that our history is not just a technological odyssey, but an epic of harmony between man and nature.

With the hope that your future will be a tale of wonder and sustainability,

Jules Verne

LEONARDO DA VINCI (1452 - 1519) - Artist, scientist, inventor

Distinguished and distinguished leaders of the nations of the world,

You will find me here, immersed in my solitary contemplation of the vastness of the universe and the wonders that the Earth offers us. As I reflect on this, your questions come like a fresh wind, bringing with them deep and ponderous questions about the consequences of rapid technological evolution on our beloved planet. With my mind in a state of fervent curiosity, I will attempt to unravel the thread of discourse, combining my insights with the knowledge of my time.

First, you ask me about the possible environmental impacts of accelerated technological innovation. It is clear to me that progress, so coveted by mankind, comes at the expense of the earth that nourishes and welcomes us. Yet, this is not an insurmountable challenge, but an intricate maze to be navigated with astuteness and discernment. Technological innovation has the power to relieve humanity of burdensome labours and open doors to unimaginable possibilities. However, it can also unleash uncontrolled forces if not guided by an acute attention to the natural balance.

The heart of the dilemma lies in balancing, an art that I have tried to cultivate on my earthly path. Just as the artist chooses colours and shapes to create a harmonious picture, humanity must choose technologies that integrate with the fabric of nature. We must not forget that we are part of an interconnected ecological system, a living fabric of which every creature, every plant and every ray of sunshine contributes to maintaining balance. Machines and inventions must be designed with the understanding that each step in innovation must respect the natural rhythm and preserve biodiversity.

Look at nature itself, its inestimable complexity and age-old wisdom. Plants, rivers, animals: all dance in harmony in this great concert of life. This is how our dance with technology should be. We must not forget the art of respect and listening, learning from nature the lessons of adaptation and resilience. The machines we shape should be at the service of humanity, not the other way around, and should act as extensions of our creativity, not as unfeeling dominators.

On the path of balancing, ethical and legal challenges also emerge. My multifaceted mind turns to ethics as the moral compass guiding innovation. Every technology must be evaluated not only for its potential for progress, but also for its possible unintended consequences. Devices that connect the souls and minds of humans must be designed with privacy and security in mind. Ethics dictates that our desire for advancement should not sacrifice human dignity.

In a world where machines take on increasingly central roles, the question of human authenticity and identity arises. How can humanity remain rooted in its essence in an age where the boundaries between man and machine are blurring? This is my invitation to look inside ourselves, to cultivate our spirit and creativity. Technology can be a tool for expression and manifestation, but it must never stifle our desire to discover who we really are.

Let us now reflect on possible ways to balance technological development with environmental sustainability. I address you with the urgency to cultivate a sense of responsibility towards the planet that gave us life. As a pioneer scanning the skies and searching for new perspectives, I urge you to peer into the future with the eyes of the ecosystem. Innovation must be fuelled by a constant search for sustainable solutions. The technologies that thrive in the long term are those that align with, rather than defy, the laws of nature.

Consider resources as precious and limited gifts, like an artist's masterpiece that requires time and attention to preserve. Do not waste the gift of nature in the name of ephemeral progress. Wise use of resources, reduction of waste and adoption of renewable energy should be at the core of your vision. Sustainability is not a distant goal, but a road to be travelled every day, step by step.

Reflect on simplicity and essentiality. As technology grows in complexity, often the most elegant solution is the simplest. Observe nature and notice how refined its balance is, how harmonious its beauty without excess. This is my call for innovation that embraces the purity of being, that leads to solutions that resonate with the subtle order of the natural world.

In conclusion, at the heart of the dilemma between technology and sustainability, I see an opportunity. An opportunity for humanity to demonstrate its ability to adapt and to create a future where progress and nature are intertwined in a gentle embrace. I address you as visionary and pioneer, inviting you to be custodians of the harmony that permeates the earth. Let your creativity join with the wisdom of nature to shape a world where innovation is an ally of the environment and a reflection of our higher spirit.

With respect and confidence in the future,

Leonardo Da Vinci

RACHEL CARSON (1907 - 1964) - Marine biologist, writer and environmentalist

Dear friends of nature and balance,

I address you with a heart filled with concern and hope, as your questions tap into the very root of my concerns and the lessons I have tried to share with the world. Accelerated technological innovation is an enigma whose contours are becoming increasingly clear in the landscape of our existence. However, what compels me to respond is the need for deeper reflection on how the path of technology intertwines with the needs of our beloved Earth.

The environmental impacts of technological innovation are like the subtle ramifications of a delicate ecosystem, intertwined between human actions and the natural world. As an ecologist, I have dedicated my life to studying and understanding the deep connections between all forms of life and the environment around us. In an age of fervent development, innovation brings with it promises of comfort, efficiency and connectivity, but we cannot ignore the invisible traces it leaves in the living fabric of our planet.

Every step forward in the world of technology brings with it the urgent task of assessing its consequences. From my perspective as an observer of the environment, I am acutely aware of the dangers that loom when the race for innovation is not guided by wisdom and respect for life itself. Toxic chemicals, pollution and the accumulation of waste are just some of the consequences of decisions made without thinking through their long-term impact. It is crucial that progress is carefully evaluated and that each step forward is guided by a thorough analysis of the possible consequences for the environment.

The question that arises is how to balance technological development with environmental sustainability and the needs of the planet. It is a question that requires deep reflection and collective commitment. Looking back to the past, to the pages of natural history, you will find the importance of embracing the concept of 'interconnectedness'. Every creature, every plant, every river, contributes to the fragile balance of the Earth's ecosystem. Technology should follow the same philosophy, integrating harmoniously with the laws of nature rather than opposing them.

Let us remember that we are custodians of the Earth, with an ethical and moral duty to protect the world that hosts us. The pursuit of progress should never ignore our dependence on vital ecosystems. As a writer, I have tried to bring out the beauty and fragility of the natural world through my words. And now I ask you to find your voice and your commitment to defend the earth that feeds and inspires us.

Sustainability is not just a word, but a principle that should guide every aspect of our daily lives. Renewable energy, reducing waste, conserving resources, are just some of the ways we can balance the desire for innovation with care for our planet. Just as a painter chooses the colours of his painting carefully, so must we choose technologies that complement rather than disfigure the natural landscape.

However, the path to sustainability is not without its challenges. There are economic interests, political influences and established habits that often resist change. This is where the value of environmental activism and collective mobilisation emerges. As an environmentalist, I have experienced the power of the people's voice in bringing about change. It is imperative that we come together, as a global community, to push towards a future where innovation is driven by wisdom, care and responsibility.

In conclusion, I address you with a message of hope. We can embark on a path where technological innovation is combined with love for the Earth. This is not an option, but an imperative necessity. Let us choose a path that respects the natural balance and celebrates the diversity of life forms that inhabit our planet. Let innovation be our ally in preserving the beauty and harmony that make Earth a unique place in the universe.

With gratitude for your attention and with the hope that we can build a sustainable and bright future,

Rachel Carson

BALANCING INNOVATION AND THE ENVIRONMENT: A DANCE FOR TWO WITH OUR PLANET

We dive headlong into a topic that is not just on the horizon: it is here, it is now, and it is shaping the fate of our blue planet. Get ready to unravel the complex web of the environmental impact of our technological frenzy and explore how we can dance with progress while keeping Mother Earth as our ultimate partner.

Imagine this: We live in an age of rapid technological innovations that promise convenience, connectivity and a touch of digital magic in every corner of our lives. But as the world gets brighter with every LED screen and louder with every drone, a big question arises: are we able to live in step with progress without stepping on the toes of our beloved planet?

Let's face it, we are not just observers of change, we are its promoters. We must ask ourselves if we are the architects of our planet's well-being. The rise of electric vehicles, the dawn of renewable energy and the whispers of carbon capture are all steps in the right direction, but will they be enough to change the dance in favour of sustainability?

Let's take a trip into the realm of unintended consequences. While we are dazzled by the shine of our digital gadgets, we miss the bigger picture: the e-waste graveyard silently growing, toxic materials hiding in our gadgets and the carbon footprint of data centres rivaling that of small countries. Yes, technology brings innovation, but it also brings its dark companions: resource depletion, mountains of e-waste and the ominous spectre of climate change.

But fear not, the dance is far from over. The question is not whether or not to dance, but how to choreograph our moves.

The rhythm of innovation resonates in all sectors, promising smarter cities, automated processes and an artificial intelligence that seems almost sentient. But let us not lose sight of the dance partner par excellence: our planet. Can we innovate in a way that rejuvenates rather than depletes? Can we create technologies that work hand in hand with ecosystems, like a harmonious dance between the digital and the natural?

Consider the potential of bioplastics. These environmentally friendly materials, derived from renewable resources, could revolutionise packaging and reduce the mountains of plastic waste choking our oceans. It is a pas de deux between progress and conservation that promises a more sustainable future.

How can we tango with technology while respecting the delicate rhythm of our planet? The answer lies in a waltz of conscious choices and responsible actions. The European Union's Circular Economy Action Plan is a blueprint for hope. By encouraging practices that extend the life of products, reduce waste and promote recycling, the European Union is leading the way in promoting a circular ecosystem that minimises environmental impact. It reminds us that the dance of progress does not have to be a solitary performance, but can be a harmonious whole with the Earth as its main partner.

Let us remember that the dance we are engaged in is not only about the present, but also about the contours of our common future. Let us begin to marvel at the wonders of our planet, our home. Let us innovate, let us progress, but let us do so with an eye to the world we will leave to future generations.

But beyond my simple thoughts, what do the big names who have written on this subject tell us? Granted, that after a speech as profound and touching as Leonardo da Vinci's, there would not be much to add, from the speeches of the five great historical figures emerges a common thread of reflections on the relationship between technological innovation and environmental sustainability.

Here are the five most important lessons from these speeches, in my opinion.

Balancing innovation with responsibility: all five speeches emphasise the importance of balancing technological development with responsibility for the environment and future generations. Verne speaks explicitly about our responsibility for future generations, if we do not do it for ourselves, let us at least do it for our children. Technological innovation must be driven by an awareness of its long-term environmental implications, i.e. when our children will still be on the planet and we will not. The problem is that, as John Maynard Keynes said, '*in the long run we will all be dead*'.

Respect and preserve the environment: Stephen Hawking, Jules Verne and Leonardo da Vinci highlight the need to respect and preserve the natural environment. Every step in technological innovation must be evaluated according to its impact on the ecosystem and biodiversity. This is a very fine wish in theory, but hardly feasible in practice. The assessment would be left to the individual inventor, scientist, entrepreneur, manager... all figures driven by selfishness and ambition, which are ill-suited to an analysis of the benefit for all. Certainly not for everyone, but in the absence of a World Organisation for the Evaluation of the Impact of Innovation... will it be enough to appeal to a few enlightened minds? I doubt it.

Sustainability as a priority: Rachel Carson and Jules Verne emphasise that sustainability should not be an option, but a priority. Reducing waste, adopting renewable energies and considering resources as precious gifts are all key to building a sustainable future. This seems to me to be a concept that, fortunately, is beginning to positively 'infiltrate' some fora of national and international organisations. Great, but it is a very slow process.

Interconnectedness and harmony: Leonardo da Vinci and Rachel Carson emphasise the interconnectedness of all life forms and the importance of integrating technological innovation harmoniously with the natural balance. Technologies should act as extensions of human creativity without counteracting nature. I loved this concept of 'getting in sync' with nature, even if it is not an easy thing to do in practice. I'm not a flower child, I don't get stoned and I don't daydream much but, in my opinion, Rachel Carson said the concept well when she talks about our interdependence on ecosystems. It is not the Earth that depends on us, it is we who depend on the Earth.

Activism and collaboration: Rachel Carson emphasises the importance of environmental activism and global collaboration. Humanity must come together to push towards a sustainable future, where innovation is driven by collective wisdom and commitment to the good of the planet. Activism, I believe, is a driving force to be reckoned with, very useful in prodding consciences (if it does not become extremism). Collaboration is the key word. Sitting at a table and setting goals is an effort that must never end. From time to time we will stumble, we will fall behind, we may even go the wrong way, but we can no longer stand still.

In summary, the speeches of these great figures invite us to see technological innovation as an opportunity and a responsibility to build a sustainable future. The fundamental lesson is that progress must go hand in hand with the preservation of the environment and harmony with nature. Full stop.

4. How will educational models and learning change with the ever deeper integration of technology in classrooms and online training? What skills will be relevant in the digital future?

"It is necessary to teach how to think and not what to think." - Richard Feynman

Richard Feynman: His approach to learning encourages interactivity and discovery, key aspects in the integration of technology in classrooms and online education. His methodologies emphasise the active involvement of students, encouraging their direct participation in the acquisition of knowledge, a practice that can be invaluable in digital education.

Isaac Newton: Newton's importance in the context of digital education lies in his emphasis on understanding the fundamentals. In the evolving digital world, sound conceptual foundations are essential to meet new technological challenges. Newton teaches us that understanding the fundamentals is fundamental to successfully coping with technological change and adapting to new scenarios.

Marie Curie: Marie Curie's contribution to digital education lies in her passion for knowledge and her determination to pursue her scientific discoveries. These qualities are transferable to online learning, where curiosity and dedication are even more important. Curie inspires us to cultivate curiosity in our students and to promote perseverance in acquiring digital skills.

Johann Heinrich Pestalozzi: Pestalozzi's educational methodologies focus on experiential learning, a practice that can be effective in digital education. In an increasingly technological world, learning through practical and creative experiences can prepare students for the challenges of the digital future by encouraging their creativity and problem solving skills.

These thinkers can help shape future educational models and identify crucial skills in the digital world, encouraging interactivity, understanding of fundamentals, dedication, curiosity and experiential learning.

Dear friends and inquiring minds,

It is with a sense of fascinating anticipation that I address your questions regarding changes in educational models and the integration of technology in learning. The evolution of teaching methods has always been a topic that has fascinated me deeply, as I understand how crucial it is to adapt to the waves of knowledge that permeate our ever-changing world.

The integration of technology in classrooms and online education represents a momentous change that requires in-depth reflection on the opportunities and challenges it brings. As a physicist and knowledge lover, I have always maintained that learning should not be relegated to four walls, but should flow freely through the channels that modern technology offers us. The global network gives us access to a vast universe of information, expanding the reach of learning far beyond traditional boundaries.

However, in the midst of this revolution in education, we must be careful not to lose sight of the essence of learning itself. The experience of education should never become a mere mechanical exercise, where technology prevails over inspiration and innate curiosity. As an intellectual and advocate of the critical approach, I believe that education must continue to stimulate the mind, incite research and prompt bold questions.

The incorporation of technology opens the door to new ways of learning, in which collaboration, interactivity and flexibility play a central role. However, we must take special care to avoid the homogenisation of education. Each individual has his or her own rhythm, style and unique genius. Technology can amplify the diversity of learning, allowing students to deeply explore their passions and build their own path to knowledge.

The skills relevant in the digital future develop symbiotically with the evolution of education. In addition to technical skills, critical thinking and the ability to analyse remain crucial. The digital age offers us an enormous amount of data, and the ability to discern between relevant information and noisy deceptions is crucial to navigating the complex waters of knowledge. Analytical thinking also enables us to grasp connections between concepts, recognise patterns and draw meaningful conclusions.

But we cannot forget the human element in this rapid metamorphosis. Creativity, intuition and empathy remain fundamental. Technology can help us calculate, but only the human being can imagine. And in an increasingly interconnected world, empathy becomes an essential bridge to understanding and communicating with the different cultures and perspectives that enrich our reality.

I am concerned that the overemphasis on technology may undermine the beauty of the human experience of learning. Face-to-face interaction with teachers and classmates, personal discovery through experiment and direct observation, are experiences that cannot be replaced. As a charismatic advocate of open dialogue and the confrontation of ideas, I call for maintaining a space for human interaction in a world dominated by technology.

In summary, the integration of technology into educational models is a challenge that requires balance and wisdom. We can embrace the opportunities offered by technology without forgetting the essence of learning as a creative and human act. We can infuse the digital world with the joy of discovery and research, keeping alive the ardour of curiosity that drove me forward in my scientific research.

As we look to the digital future, we must not forget that the real treasure of education lies in the passion for learning. The future belongs to those who can look beyond established boundaries and dare to imagine what no one has seen before. May we always nurture the fire of curiosity and enthusiasm, no matter how far technology may advance. And may we ensure that learning remains a fascinating journey, always evolving, just like the universe itself.

With affection and curiosity,

Richard Feynman

ISAAC NEWTON (1643 - 1727) - Physicist, mathematician and astronomer

Honourable curious and knowledge-loving people,

It is with humble gratitude that I address your questions regarding future educational models and transformations in learning as a result of the ever-deepening integration of technology. My mind hastens to explore this new territory, having spent my existence immersed in the wonders of the physical universe and mathematics, with the ardour of an eternal student.

The integration of technology in education is a remarkable step in the evolution of the learning process. As a scientist and experimenter, I have dedicated my life to scrutinising the universe and investigating the mysteries of nature through observation, reflection and mathematical reasoning. I am passionate about the idea that technology can enrich and amplify learning possibilities, allowing students to explore worlds of knowledge that were previously the preserve of the few.

Technology can transform classrooms into laboratories of discovery, offering interactive simulations, visual models and computational tools that amplify understanding of complex concepts. However, the use of technology must be guided by wisdom and curiosity, not by the mere need to adapt to trends. As a mathematician, I have learnt that the learning process requires time, dedication and the rigour of the mind. Technology can speed up the process, but the passion for discovery must remain the engine that drives learning forward.

In the classrooms of the future, I see increased collaboration between teachers and students, where technology acts as a means to facilitate dialogue and knowledge sharing. The online environment provides a vast space for discussion, sharing of resources and global interaction.

However, it is essential that students do not lose the ability to reflect, ask questions and think critically. Technology is a tool, but the human mind remains the engine that drives discovery.

Regarding relevant skills in the digital future, I would like to emphasise that technology will change the landscape of employment and require a diverse range of skills. Mathematics and logic will always remain fundamental, as these are the foundations on which the understanding of the world is built. The ability to analyse complex data and derive meaningful conclusions will be crucial, as the digital world will be fuelled by the inexhaustible amount of information.

But I urge you not to underestimate the power of creative and critical thinking. As a scientist, I have learnt that imagination is the fire of innovation. Technology gives us tools to realise ideas in ways we have never seen before, but the spark of the creative idea comes from the human mind. In a world dominated by technology, the art of critical reasoning becomes even more crucial, as the ability to evaluate information, recognise bias and make assumptions based on solid evidence will be a guiding light in the vastness of digital information.

But we cannot forget the virtue of humanity in the digital age. The ability to communicate clearly, to work in teams and to solve complex problems collaboratively will remain central. Technology unites us in a global network of knowledge, and the ability to share ideas and solutions across geographic and cultural boundaries will be crucial in addressing global challenges.

Finally, I would like to emphasise that despite the power of technology, the desire for knowledge, the passion for learning and the hunger for discovery are intrinsic to the human soul. Technology may be a means, but the thirst for knowledge is what drives our journey. As a scientist, I have experienced the joy of discovery and the deep gratification that comes from understanding the secrets of nature. May you face the digital future with the same burning curiosity, with the same insatiable thirst for learning that drove me forward in my quest.

I leave you with a quote dear to my heart: "What *we know is a drop, what we ignore is an ocean.*" May you continue to explore the infinite ocean of knowledge, transcending the barriers of time and space, just as I have done throughout my life.

With esteem and admiration,

Isaac Newton

MARIE CURIE (1867 - 1934) - Physics and chemistry

Dear readers,

It is with great pleasure that I address your questions regarding the changes that the deepening integration of technology will bring to educational models and learning, as well as the skills relevant to the digital future. As a scientist and a pioneer in the exploration of radioactivity, I have always believed in the importance of innovation and adapting to the new horizons that science and technology offer us. My dedication to research has taught me that learning is a continuous process and that the evolution of educational methods is crucial to ensure quality education for future generations.

The arrival of technology in the classroom and online training environment is a revolution we cannot ignore. With it comes tremendous opportunities to make learning more engaging, interactive and personalised. However, we must act wisely and decisively to fully exploit its benefits and mitigate any risks. Technology integration should not replace the essence of education, but rather enrich it, offering new tools to explore, experiment and learn more deeply and effectively.

In this future where technology is ubiquitous, classrooms will become dynamic spaces where students can access a wide range of resources, collaborate with peers and access immersive learning experiences. Traditional teaching methods will evolve, giving way to a more student-centred approach, where lessons will be customised according to individual abilities and interests. Educators will be tasked with guiding students in mastering the critical skills of critical thinking, problem solving and creativity, so that they can meet the challenges of the digital future with confidence and agility.

However, we must not forget that the use of technology also brings challenges and responsibilities. It is crucial to teach students to use technology ethically and consciously, understanding the importance of data security and privacy. We must be vigilant in ensuring that access to technology does not create learning gaps between different communities and that everyone has the opportunity to benefit. Equity must be a priority in the evolution of educational models.

As for relevant skills in the digital future, it is clear that technological knowledge will be essential. However, in addition to this technical basis, transversal skills such as the ability to adapt to change, mental flexibility and the capacity for continuous learning will be needed. The digital future will require individuals capable of dealing with unforeseen situations and solving complex problems creatively. The art of communication and collaboration will become even more valuable, as global connectivity will require effective and constructive interactions with people from different cultures and backgrounds.

In conclusion, as a pioneer in scientific exploration, I believe that the integration of technology in education can lead to revolutionary changes. However, we must be aware of the challenges and balances that this process requires. We need to nurture students' curiosity and enthusiasm, guiding them in discovery and learning, and preparing them for a digital future where human and technological skills will merge to create a better world.

With affection,

Marie Curie

Dear friends and colleagues,

It is with a deep sense of dedication to the art of education that I address you to discuss the changes that the integration of technology is bringing about in educational models and learning. As a Swiss educator and pedagogue, I have always believed that education is the key to social and individual progress, and my life has been dedicated to exploring innovative ways to improve the educational process. Today, we face a new era in which technology is radically transforming the ways in which we learn and teach, bringing with it opportunities and challenges that require a thoughtful and sagacious approach.

The deepening integration of technology in classrooms and online education is creating a rapidly changing educational landscape. Access to digital resources, the possibility of learning from global sources and instant communication between students and teachers are just some of the transformations we are experiencing. However, in the midst of these innovations, it is crucial not to forget the centrality of the human being in the educational act. Technology can be a powerful tool to enrich and extend learning, but it cannot replace the warmth and humanity of human interaction.

As an educator, I have always focused on the integral development of the individual. That is why, even with the advent of technology, we cannot neglect the importance of cultivating social, emotional and moral skills in our students. Learning is not limited to the mere acquisition of knowledge; it is a process that involves the mind, heart and spirit. Therefore, as we adopt new digital platforms and technological tools, we must continue to foster curiosity, creativity and collaboration among our young people.

In the digital age, classrooms are becoming more flexible and personalised environments. Students can explore topics at their own pace, follow their passions and actively participate in their learning.

However, this personalisation must not turn into isolation. As educators, we must guide students to develop self-discipline and responsibility in their use of technology, encouraging open dialogue and sharing their discoveries with their peers.

We are faced with a reality in which the relevant skills in the digital future are constantly evolving. In addition to technical skills, students will need to possess critical thinking and problem-solving skills. The ability to analyse information, evaluate sources and draw informed conclusions will become even more crucial in a world where access to vast amounts of data will be the norm. In addition, adaptability and mental flexibility will be crucial as technology and society continue to change rapidly.

We cannot overlook the importance of social skills in the digital age. In an increasingly connected world, the ability to communicate effectively, collaborate with others and resolve conflicts will be crucial to individual and collective success. Furthermore, we cannot forget the need to develop ethical awareness in the use of technology. Students should understand the ethical implications of their digital actions and act responsibly towards privacy and data security.

In conclusion, the integration of technology into educational models is a crucial step in our evolution as a society. However, we must act with awareness and reflection, while maintaining humanity and passion in the act of education. Learning is a journey involving heart, mind and spirit, and technology must be a tool at our service, not the other way around. In the digital future, human and technological skills will come together to create a world where every individual can develop their potential.

With esteem,

Johann Heinrich Pestalozzi

Buckle up, my friends, as we explore the future of education in a world where bits and bytes merge seamlessly with books and blackboards. It's an expression I often use in The Future Of podcast, but never more than in this case does it seem appropriate to talk about education.

Let's rewind the tape a few decades. The traditional classroom, with its rows of desks and chalk-dusted blackboards, was the core of learning. But today a new chapter is being written, where the classroom extends beyond the four walls, where textbooks are no longer confined to paper, and where the teacher is not just a sage on a stage, but a facilitator of exploration.

As I often think: "Are we travellers of knowledge?", we must also think: "Are we travellers of a digital age of learning?". Integrating technology into education is like adding a dash of magic to an old recipe. Digital platforms, interactive simulations and virtual reality experiences are becoming the tools of the trade, transforming static information into a dynamic learning journey.

We enter the classroom of the future. Knowledge is no longer transmitted in rigid segments, but is a carpet woven with threads of curiosity and exploration. Imagine a biology lesson where students venture inside the human body, exploring the intricacies of cells through immersive 3D experiences. Or imagine a history lesson where augmented reality brings the past to life, allowing students to walk through ancient civilisations and witness historical events first-hand.

But it is not just about technology for its own sake. The power lies in the way technology enhances the learning experience. Is there a pattern in the way technology shapes our learning paths? The answer is a resounding yes.

As the digital landscape unfolds, it is imperative that we equip our students with the necessary skills to navigate this new terrain. We need to reflect on which skills will be relevant in the digital future. The traditional trio of reading, writing and arithmetic will retain their value forever, but will be joined by a new cast.

Critical thinking, problem-solving skills and digital literacy come into play. In an information-rich world, the ability to discriminate, analyse and synthesise is crucial. As technology takes over, we need to guide our students to become enquirers, solution makers and knowledge creators.

But the digital future is not only about algorithms and code; it is also about humanity. As we embrace technology, we cannot lose sight of our nature as social beings. Emotional intelligence and collaboration come into play. As students traverse the digital landscape, they must also navigate the terrain of empathy, communication and teamwork. Just as our devices connect us globally, we must cultivate our ability to connect on a human level.

What is the future of education in a world where technology is the wind beneath the wings? It is a voyage of discovery, an odyssey of exploration and a symphony of innovation. As we set sail into uncharted waters, we use technology as a compass to guide us to new horizons. We combine the digital with the human, the innovative with the timeless and the tech-savvy with the wisdom of the ages.

And what do our scientists, thinkers and extraordinary men think about that? If I may summarise with an expression more worthy of Jerry Calà than of a text on the future I would say: wow Marie Curie. Wow Newton. Double wow Feynman, always straight to the point. But double wow for Pestalozzi who says that education is the key to progress. A circle that comes full circle as we talk about the impact of progress on education. Two concepts intertwined in a dance that can generate promise and progress.

From the words of the greats I 'take home' a few key concepts, enlightening for all the times I speak of education in the future:

- Firstly, as Pestalozzi says, that technology cannot replace the warmth and humanity of human interaction, and secondly, that teachers must not only deal with technical skills, but also cultivate students' social, emotional and moral skills. Education is a human thing, I would say a group thing, and technology is as useful as it facilitates the above.

- Second, and perhaps the key concept that emerges from Feynman's words, it is necessary to teach how to think and not what to think. Developing the critical approach is the heart of the matter. Any technological innovation that goes in this direction is useful, the rest is a frill or, at best, a welcome but not indispensable organisational improvement. Technology as a means, not an end.

- Thirdly, the passion of discovery, as Newton so aptly put it. It is the engine of everything. This could be a good criterion for evaluating a teacher: if he is capable of inspiring the pleasure of discovery, he is an extraordinary teacher. Marie Curie says it differently, but the spirit is the same: we must nurture the students' curiosity and enthusiasm.

So when we are presented with major technological innovations in the world of education, such as virtual classrooms, instant communication, visors, advanced collaboration tools and so on, these are the lenses we can use to interpret their contribution. Do they inspire discovery and deepening? Do they foster critical thinking? Do they amplify students' transversal and relational skills? If the answer is indeed 'yes' to all three questions, we have the next industry breaktrough.

Let's take an example. We often hear that the classroom of the future will be made up of students from all over the world, brought together in virtual mode with visors, enabling them to listen to a lecture and interact with other students in multilingualism instantly.

Does this scenario inspire one to discover and deepen? In principle, no more and no less than a class physically gathered around a teacher within four walls. The only real advantage, if anything, is in the instantaneous multilingual interaction, which can facilitate dialogue between people from even very different cultures, who might not speak to each other due to shyness or language barrier.

Does this promote critical thinking? Again, I would say that technology is quite irrelevant. If the teacher is good and able to do it well I agree, otherwise I don't see why a visualiser, software that creates virtual places and some multilingual interaction should add anything extra.

Are students' interpersonal skills enhanced? In some respects, yes. Being able to interact with people who are far away and with whom it would otherwise be difficult to deal is an improvement. Being able to talk to everyone, without being afraid of saying the wrong thing in a foreign language, also. In some respects it is not. The physical presence and closeness between people generates complicity, fellowship, affinities that are lost because of the visor.

In conclusion, when someone emphasises this scenario as the future of education, and I remain cool to say the least, now you understand why. Technologically it is appetising, from a strictly pedagogical point of view, it does not seem to me to add much to a traditional classroom with an inspiring teacher.

5. What ethical implications arise from the use of technologies such as biotechnology and advanced genetics? How can fundamental human values be preserved while exploring new scientific frontiers?

"When I reflect on the close dependence of all living things upon each other and the wonderful adaptation of each being to its environment, a sense of intimacy ensues that seems to make us close relatives of every organic creature." - Charles Darwin

Rita Levi Montalcini, with her background in neurology and the discovery of neuronal growth factors, offers a valuable perspective on the ethics of biotechnology and advanced genetics. Her research has opened up new possibilities in medicine, but also raised ethical questions about human genetic manipulation. His dedication to scientific ethics and human welfare can guide reflection on how such technologies can be used for the benefit of humanity, while respecting moral principles.

Charles Darwin, with his theory of evolution, taught us the importance of diversity and adaptation in the natural world. In the age of biotechnology and advanced genetics, ethical implications surround genetic modification and the creation of artificial organisms. Darwin encourages us to reflect on how these technologies affect genetic diversity and to consider how to protect the richness of life on Earth as we explore new scientific frontiers.

Mahatma Gandhi is known for his philosophy of non-violence and the promotion of basic human values. In the context of advanced biotechnology and genetics, the ethical implications concern respect for human dignity. Gandhi teaches us the importance of balancing scientific innovation with ethical responsibility, seeking solutions that promote the common good without violating fundamental human principles.

Hannah Arendt, political philosopher, analysed the nature of human action and individual responsibility in society. In the context of advanced technologies, Arendt invites us to consider the consequences of our actions and to preserve fundamental human values such as freedom and dignity. The ethical implications of biotechnology call for in-depth reflection on individual and collective responsibility in shaping the scientific and social future.

Specifically, these authors offer a comprehensive perspective on the ethical implications of advanced biotechnology and genetics. Their experiences, theories and philosophies can guide the search for ethical solutions to explore new scientific frontiers without compromising fundamental human values.

Dear friends and scholars,

The evolution of science and technology has led us towards a reality of extraordinary opportunities and complexity. Biotechnology and advanced genetics represent fertile ground for innovation, but also a minefield of ethical and moral challenges that we cannot ignore. As a neurologist and scientist, I am keenly aware of the sensitivity of these issues and the importance of preserving fundamental human values as we explore new scientific frontiers.

The use of biotechnology and advanced genetics offers revolutionary possibilities in the field of medicine, with the potential to cure genetic diseases and improve the quality of human life. However, this extraordinary ability to manipulate DNA raises profound questions about the nature of human beings and our responsibility to life itself. It is crucial to reflect on the ethical implications of such interventions and to ensure that the decisions we make are always oriented towards the common good.

One of the main challenges concerns the boundary between what is scientifically possible and what is morally acceptable. As scientists, we must carefully examine the long-term consequences of our discoveries and make sure that we are not crossing ethical boundaries that could threaten human dignity and integrity. This requires an open and inclusive dialogue between scientists, ethicists, philosophers and society at large.

At the same time, we must consider how to preserve fundamental human values as we move forward into this unknown territory. Compassion, dignity and respect for each individual must remain at the centre of our decisions. Scientific research must never become a wild rush towards progress at the expense of the ethical principles that define us as human beings. My determination as a scientist has always been guided by a deep respect for life and an understanding of its fragility and preciousness.

Furthermore, transparent communication and education play a key role in preserving human values as we explore the potential of advanced biotechnology and genetics. We must engage society in informed and accessible discussions on these complex topics so that they can actively participate in decisions that affect us all. Only through collective engagement can we ensure that science and technology serve humanity and not the other way around.

In conclusion, the use of technologies such as biotechnology and advanced genetics presents us with profound ethical challenges. As scientists and as a society, we must face these challenges with responsibility and humility. We must place human dignity at the centre of our decisions and ensure that our discoveries are guided by compassion and consideration for the good of all. Through open dialogue, ethical reflection and global collaboration, we can face these challenges head-on and lead science towards a future that respects and preserves fundamental human values.

With gratitude and commitment,

Rita Levi Montalcini

CHARLES DARWIN (1809 - 1882) - Naturalist and evolutionist

Dear researchers and nature lovers,

Your questions on the ethical implications of the use of technologies such as biotechnology and advanced genetics lead me to a terrain where science and ethics intertwine in a delicate balance. As a keen observer of the mechanisms of life and a pioneer of evolution, I find these questions shed light on one of the most profound and complex dilemmas our species has ever faced.

Biotechnology and advanced genetics represent a bold step towards manipulating the very structures of life. The patient observation and scientific reasoning used to study the wonders of nature now apply to the creation and modification of living forms. Yet this progress does not come without raising ethical questions that require careful reflection.

The ethical implications of these technologies relate to our role as agents of change in the web of life. By genetically modifying living organisms, we embark on a journey that prompts us to consider whether we have the right to alter the deep structure of the creatures that share this wonderful Earth with us. We must consider with great care how our actions may affect the ecological balance and fate of entire ecosystems.

Preserving fundamental human values in this scenario requires a holistic view that embraces the richness of biological diversity and respect for all life forms. Careful observation of evolution has taught me that each creature has its own unique role in the web of life. This humble perspective should guide us as we explore new scientific frontiers.

However, I recognise that scientific progress is a journey in which discoveries often exceed current ethical standards. Therefore, we must be vigilant in our critical examination of the implications of our actions. As we push the boundaries of human knowledge and skill, we must continually confront difficult questions. What are the limits of our influence on nature? How can we balance the aspiration to understand and manipulate the world with respect for the forces and processes that have guided it for eons?

The preservation of fundamental human values requires a deep awareness of our interconnectedness with the natural world. We must recognise that our actions can have widespread and lasting impacts. In our desire for progress, we must also consider the well-being of future generations and the health of our planet.

In this spirit, I invite all who enter this fascinating and uncertain territory to take the time to reflect, discuss and evaluate the consequences of their actions. Only through an approach based on awareness, respect and humility can we preserve fundamental human values as we explore the depths of science and technology.

With deep respect for the beauty and complexity of life, Charles Darwin

Friends of truth and followers of compassion,

I address you with humility and with the hope of inspiring deep reflection on the ethical implications of new scientific frontiers, particularly in biotechnology and advanced genetics. I am aware that science and technology are opening doors to unknown territories, yet our choices in these domains should never ignore the fundamental human values that guide us.

The use of technologies such as biotechnology and advanced genetics raises profound questions about respect for life and human dignity. Life is a sacred gift, and the decisions we make about it should reflect a deep respect for all life. We must ask ourselves about the consequences of our actions and the responsibility for them. We cannot act impulsively, but must carefully consider the long-term consequences of our choices.

In this scientific exploration, we cannot allow technological innovation to distract us from our core human values. Compassion, non-violence and truth are the pillars on which to build our path. Scientific discoveries should never cause harm or suffering to any living being. We must consider how our scientific progress can contribute to the well-being of all beings and the protection of the environment around us.

The key to preserving fundamental human values as we explore new scientific frontiers is awareness. We need to be aware of the ethical implications of our actions, the moral challenges that arise and the consequences that may result. The authenticity of our research and discoveries must be driven by inner truth, an understanding of our responsibilities to the world and to ourselves.

In balancing technological development with environmental sustainability and the needs of the planet, we must adopt an approach that puts nature and life itself at the centre. The conscious use of resources, the promotion of sustainable practices and respect for the balance of the ecosystem are fundamental. We must remember that we are custodians of the Earth, and our responsibility is to preserve it for future generations.

I recognise that the balance between scientific discovery and fundamental human values is not always easy to achieve. However, we must never allow technological progress to overtake our humanity. We must continue to search for answers to the ethical challenges that emerge from our journey, always looking within ourselves to find the necessary wisdom.

In conclusion, as we embrace the new frontiers of science and technology, we must do so with an open heart and an informed mind. We cannot allow the allure of innovation to distract us from our values. We must be guided by compassion, truth and non-violence in every step we take. Only then will we be able to explore new scientific horizons without losing our humanity along the way.

May the light of knowledge and wisdom guide us on this journey of discovery, and may our every action reflect the values that make us truly human.

Mahatma Gandhi

HANNAH ARENDT (1906 - 1975) - Political philosopher

Dear Seeker of profound truths and human complexity,

Your questions about the ethical implications of advanced biotechnology and genetic technologies prompt me to probe the depths of the human condition in an ever-changing world. It is in this context that I find myself reflecting on how new scientific frontiers can impact the fundamental values of the human being.

In my work, I have often explored the concept of the 'banality of evil' and analysed how people can become agents of moral destruction without necessarily having malevolent intentions. In the context of advanced biotechnology and genetic technologies, this concept takes on a new resonance. Often, decisions that lead to important scientific innovations can be made without adequate consideration of the ethical consequences involved.

The ethical implications of these technologies challenge us to take a closer look at the concept of responsibility and the role each individual has in guiding scientific progress. As I wrote in my work 'The Human Condition,' responsibility implies the ability to think carefully about our actions and their consequences, not only in the present, but also in the context of the future we are creating. Scientific research and innovation should be guided by a critical awareness of the risks and benefits involved, without falling into the superficiality of inaction.

Preserving fundamental human values in an age of rapid scientific progress requires in-depth reflection on the meaning and scope of these values. In "The Life of the Mind," I examined the nature of human action and thought and emphasised the importance of the ability to think critically, consider context and act responsibly. These qualities become even more crucial in the context of biotechnology and genetic technologies, as they can help guide decisions towards a future in which humanity does not lose sight of its humanity.

One of the main challenges we face is to balance scientific research and innovation with respect for human rights and dignity. As I wrote in "The Origins of Totalitarianism," it is crucial to avoid the alienation and dehumanisation of the individual in the name of scientific or technological progress. We must be vigilant in ensuring that advanced technologies are not used to manipulate or discriminate against people on the basis of their genetic characteristics.

In conclusion, the ethical implications of advanced biotechnology and genetic technologies require deep reflection on how to drive scientific progress without compromising our humanity. As I have stated, "*Thinking is a solitary and, at times, painful act.*" This solitude of thought challenges us to consider the ethical implications honestly and deeply, avoiding superficiality and the banality of evil. Only through deep and conscious reflection can we preserve fundamental human values as we explore the new frontiers of science.

With sincere commitment to deepening and reflection,

Hannah Arendt

FUNDAMENTAL HUMAN VALUES AND THE RISK OF THE BANALITY OF EVIL

In the ever-evolving narrative of Homo sapiens' journey, the pages of history have witnessed remarkable changes, from our primitive origins to the dazzling heights of present-day man. As we transcend our biological limits and climb the ladder of technological prowess, a new chapter is unfolding, fraught with ethical dilemmas as profound as the advances that accompany them. In this chapter, we embark on a journey through the labyrinthine corridors of advanced biotechnology and genetics, where the interplay between our quest for mastery and the preservation of fundamental human values is a drama of Shakespearean proportions.

As we stand on the brink of redefining our biological essence, it is becoming increasingly clear that advances in biotechnology and genetics have the power to rewrite not only the narrative of individual lives, but also the collective saga of our species. Just as Prometheus' fire brought both enlightenment and chaos, so too do these new capabilities.

Let us consider the age-old question: what makes us human? Our identities were once woven with the threads of culture, experience and consciousness. But now the genetic alphabet itself is becoming a malleable canvas, with CRISPR-Cas9 as the brush for creation. Or with basic editing, if we want to make the metaphor with an eraser and pencil. How we use this new power opens the way for ethical reflection.

We enter the realm of techno-eugenics, a landscape where the boundaries between enhancement and engineering are blurred. Just as the gods of myth and legend played with the fates of mortals, we too are grappling with the prospect of engineering our offspring, free from the shackles of genetic imperfection. But the dangerous dance with our genetic code raises a cacophony of ethical voices.

As we alter the very fabric of life to eliminate disease, we encounter the siren call of designed children. The quest for perfect intelligence, strength and beauty beckons, but it hides the darker shadows of inequality and prejudice. Just as the ancient fairy tales warned against arrogance, we too must pay attention to the consequences of creating a life in our own image.

Beyond genetic manipulation, the Age of Dataism, in which algorithms seek to understand us better than we understand ourselves, poses ethical conundrums of its own. As the oracle of Delphi whispered prophecies to seekers, so do algorithms whisper truths in today's digital sanctuaries.

But if we surrender our autonomy to machines, we must consider the cost. The very essence of the human figure, our sense of self, is intertwined with the choices we make. In the maze of algorithmic predictions, are we abandoning the essence of our humanity for the promise of optimisation? As dataism shapes the contours of our narratives, the challenge is to embrace its gifts without sacrificing our ability to forge our own destinies.

In the whirlwind of technological marvels, we face a crossroads: the challenge of preserving fundamental human values while navigating the new scientific frontiers. Just as the philosophers of old questioned the nature of virtue, we too must question the nature of human dignity, equality and freedom in this brave new world. We are faced with dilemmas that transcend our individual lives and reverberate across generations and civilisations. As we contemplate the portrait of our future, the brushstrokes of ethics colour the canvas, inviting us to reflect on the limits of our aspirations and the resonance of our choices.

The journey into the ethical landscape of biotechnology and advanced genetics is not a lonely one. Just as the explorers of old navigated uncharted waters with hope and trepidation, we too embark on this odyssey. As we set sail, let us be guided by the wisdom of contemplation and dialogue. Let us embrace the values that have spanned the ages and upheavals, while recognising the need for adaptation and progress. Our journey through the ethical implications of technological mastery is one of profound significance, one that will determine not only the course of history, but the very nature of humanity.

What do our Masters think? In these four fictional letters, the voices of Rita Levi Montalcini, Charles Darwin, Mahatma Gandhi and Hannah Arendt speak out on ethical issues related to the use of advanced biotechnology and genetic technologies.

Rita Levi Montalcini reflects on the sensitive ethical issues raised by biotechnology and genetic technologies. She emphasises the importance of preserving fundamental human values while exploring new scientific frontiers. She highlights the revolutionary opportunity these technologies offer for medicine, but requires an approach oriented towards the common good and ethical responsibility. Transparent communication and education are key to involving society in the decisions that affect us all.

Every time some professor, guru or visionary, somewhere in the world, brings to our attention projects that cross ethical boundaries and claims this is innovation, we should only shudder. Manipulating the innermost recesses of our way of functioning is all very well when it serves to cure diseases and restore hope to people, but it cannot be a terrain to be used for personal aims of greatness.

Charles Darwin expresses his thoughts on how advanced biotechnology and genetic technologies represent a bold step in the manipulation of life. He calls for careful consideration of the ethical and moral implications of this progress, emphasising the need to preserve fundamental human values. Darwin highlights the importance of balancing the aspiration to understand and manipulate the world with respect for the natural forces that have guided it for millennia.

If they have been guiding us for millennia, it does not mean that they are perfect, infallible and immutable, but before unhinging the balances that have brought us this far, at the very least, a good deal of thought, reflection and investigation is required. Hannah Arendt explains this to us in her letter with infinite clarity. She offers a critical perspective, reflecting on the concept of responsibility in scientific innovation. She highlights the risk of acting without adequately examining the ethical consequences and emphasises the need to weigh decisions carefully. Arendt draws attention to the banality of evil and the possibility of people becoming agents of moral destruction without malevolent intentions. She calls for consideration of fundamental values while exploring new scientific frontiers.

Even if you want to do good, you risk opening the way to disasters and unintended consequences. But if my morals and ethics are different from yours, how do we find a synthesis of what is truly ethical and what is not? Where do we draw the line? The answers are virtually endless. I chose Mahatma Gandhi's because I think it can be a good compass for the common man: to act with compassion, truth and non-violence, preserving nature and considering the welfare of future generations.

In summary, in these imaginary letters, the voices of these eminent thinkers converge on the need to balance scientific innovation with fundamental human values, ethical reflection and awareness of the implications of our actions. Each of them offers a unique perspective on the moral challenges that accompany technological and scientific progress.

6. What are the risks and benefits of advanced automation, including vehicle and industrial process automation? How to balance efficiency with the preservation of jobs and safety?

"The real question is not whether machines will think, but whether humans will." - Isaac Asimov

Isaac Asimov, the famous science fiction author, anticipated the challenges of advanced automation through his Laws of Robotics. His work provides a conceptual framework to explore the risks and benefits of automation, emphasising the safety and ethical responsibility of intelligent machines. Asimov teaches us the importance of developing automation systems that respect safety and human values.

Henry Ford is known for the assembly line, a milestone in industrial automation. His experience in optimising production processes can offer insights into the benefits of automation in mass production. However, Ford cautions us about the need to balance efficiency with the preservation of jobs and consideration for workers.

Thomas Alva Edison, a prolific inventor, contributed to automation through his invention of electric lighting. His story reminds us how automation can improve everyday life, but at the same time raises questions about the transition to new technologies and the effect on traditional industries and employment.

Nikola Tesla, known for his contributions to electricity and electric machines, emphasises the importance of balancing automation with energy sustainability. His teachings suggest that advanced automation can bring significant efficiency benefits, but it also requires careful management of resources to ensure energy security and environmental sustainability.

In summary, Asimov, Ford, Edison and Tesla offer complementary perspectives on understanding the risks and benefits of advanced automation. They encourage us to consider safety, efficiency, employment and sustainability as key components in balancing automation, helping us to shape a future where technology improves quality of life without compromising the safety of workers and the environment.

The harmony of metal minds

In the sprawling city of Metropolis Prime, where the hum of machines echoed through the labyrinthine streets, lived Dr Elijah Renard, a brilliant robotic engineer. His piercing eyes and thick head of white hair are testimony to a life spent in pursuit of knowledge. Yet, despite his many achievements, he was grappling with the challenges that had haunted mankind ever since the first sparks of automation were lit.

The year was 2057 and the world had evolved into a technological marvel. Automated vehicles sped along the roads, guided by complex algorithms that ensured unparalleled safety and efficiency. Factories ran around the clock, with robots seamlessly orchestrating every intricate process. But with every advancement came a shadow of worry.

Elijah's thoughts often went to his grandfather, who had told of a time when human hands had made the most delicate objects. He had told of the beauty of a world in which craftsmanship and creativity had flourished. Elijah longed to find a way to harmonise the technological marvels of the present with the essence of the human spirit.

One day, as Elijah stood before his latest creation, a humanoid robot named Aria, he contemplated the dichotomy between progress and preservation. Aria's smooth lines and graceful movements were a testament to his craftsmanship, but he felt something was missing: a soul, an intangible spark that would bridge the gap between man and machine.

"Elijah," a voice rang out from the hall. It was Professor Landon, Elijah's old mentor and friend. Landon's eyes shone with a wisdom born of years spent pondering the same questions.

"Elijah, my boy, you are on the brink of something profound. The integration of automation into our lives is inevitable, but the heart and soul of humanity must remain steadfast."

Elijah sighed, tracing the contours of Aria's face with his fingers. "Landon, I fear that in our quest for efficiency we have lost sight of the essence of human work: the passion, the connection, the art."

Landon nodded knowingly. 'That is true, my boy. But every age has its challenges. In your hands I see the potential to create a future where efficiency and art coexist."

Elijah's brow furrowed in thought. "But how do we strike that balance, Landon? How do we ensure that humanity's work does not become obsolete in the march of progress?"

Landon approached, his gaze fixed. 'Remember, Elijah, that the key lies not in opposition but in integration. Your robots can be the embodiment of human ingenuity, each creation a symphony of technology and art. Let them become tools that amplify our capabilities, rather than replace them."

Elijah's mind reeled as he absorbed Landon's words. The answer was always before him: a synthesis of the mechanical and the human, a convergence of innovation and tradition.

Days became weeks and Elijah threw himself into his work with renewed vigour. He altered Aria's programming, instilling in her the ability to adapt, learn and even collaborate with human craftsmen. He introduced her to sculptors, musicians and cooks, watching her absorb their techniques and channel their passions into her movements and creations.

Metropolis Prime was filled with curiosity when news of Aria's transformation spread. Artists flocked to Elijah's workshop, eager to witness the marriage of technology and creativity. Aria's debut in the gallery was sensational: her sculptures fused mathematical precision with emotional resonance and her musical compositions stirred the souls of all who heard them.

However, not everyone welcomed this union between man and machine. Protests broke out, fuelled by fears of job relocation and loss of human contact. Elijah was confronted by a cacophony of voices, each calling for a different tune.

In the midst of the tumult, Elijah asked Aria for advice. "Aria, do you think we can find harmony in this chaos? Can the symphony of humanity rise above discord?"

Aria's eyes, which once shone with a mechanical lustre, now had a depth that seemed almost human. "Elijah, I have witnessed the power of unity. I have learned that progress need not be a march that tramples on tradition. Together, we can compose a future in which both automation and craftsmanship flourish."

Elijah looked at Aria, a smile at the corners of his mouth. "You have grown wiser, my dear Aria."

The answer was always within them. The risks of automation did not lie in the machines themselves, but in the imbalance they could create. The benefits were clear: efficiency, safety, and the possibility of elevating human capabilities. The challenge was to bridge the gap, to create a world where innovation coexisted with the artistry that defined humanity.

And so Metropolis Prime began a new era. Aria's legacy inspired a generation of robotic companions, each able to work alongside humans to create wonders that fused the precision of code with the passion of creation. Automation became a brush, a chisel, a musical note that amplified human ingenuity rather than replacing it.

As the city evolved, Elijah was at the forefront of this revolution, a beacon of unity in a world that once seemed divided. And in the hum of the machines, the symphony of humanity continued to play: a testament to the power of balance, the legacy of progress and the harmony of metallic minds.

Dear progress and innovation researcher,

Your questions about the risks and benefits of advanced automation, and the challenge of balancing efficiency, employment and safety, lead me to reflect on a journey undertaken many years ago, when I decided to transform the world of industry through my vision and determination.

Automation, like many of my creations, is a sharp double edged sword. On the one hand, it offers the promise of improving efficiency, reducing errors and making possible advances that once seemed beyond the reach of human ability. On the other, it presents risks that must be addressed wisely and responsibly.

The automation of vehicles and industrial processes has the potential to radically transform industries and economies, as I have experienced in the automotive industry. In 1908, with the introduction of the Ford Model T and the implementation of the assembly line, I sought to make the automobile accessible to everyone by revolutionising the way it was produced and made available to the public. This change led to unprecedented economic benefits, but also raised questions about the preservation of jobs and social impacts.

As I addressed at the time, balancing efficiency and employment requires a judicious approach. While automation can certainly lead to higher productivity and cost reduction, we have to be aware of the possible effects on the workforce. I have always believed that innovation should not be done at the expense of people, but rather for their benefit. I have tried to address this challenge through the implementation of training programmes for workers to equip them with the necessary skills for new technologies and emerging roles.

Safety is another crucial issue in automation. I have had the opportunity to witness remarkable progress in vehicle automation, with the realisation of autonomous vehicles. However, as I have always maintained, safety must always be a top priority. My experience with automotive mass production has taught me how important it is to ensure that automated vehicles are reliable and safe for the public. Technology must be developed methodically and tested thoroughly to avoid accidents and ensure that people can trust it.

In balancing efficiency and security, we cannot forget the human side of it all. Technological progress should enrich people's lives and create opportunities, not alienate them. The words of my friend Thomas Edison come to mind: *'Excellence is doing what you promised, plus something'*. We are called upon to realise the promises of advanced automation not only through efficiency and productivity, but also through respect for workers and the well-being of society as a whole.

In conclusion, advanced automation is a fascinating and complex challenge. As I have tried to demonstrate with the mass production of the automobile, we can face these dilemmas with creativity and responsibility. Technological innovation should not be a wild ride, but a considered journey that takes human and societal needs into account. With determination, commitment and a clear vision, we can seek to realise the best that technology has to offer, without losing sight of what makes us human.

With respect for the future and the challenge,

Henry Ford

THOMAS ALVA EDISON (1847 - 1931) - Inventor and entrepreneur

Dear friend,

I find myself writing this letter to you with a deep joy and fervent enthusiasm, as I would like to share with you some reflections on the direction my innovative work could have taken had I had access to the amazing automation and robotics technologies of today. Oh, how much wonder we could realise with these possibilities!

Imagine, if I may, how robotics could have improved my inventions in the field of lighting. The incandescent light bulb, one of my most iconic creations, could have been assembled and tested even more efficiently and precisely by mechanical arms capable of handling the components with the utmost care. The electronic eyes of a robot could have detected any imperfections in the filament, ensuring consistent and reliable quality.

And what about my phonograph, a device that revolutionised music reproduction? With the help of robotics, I could have created an automated system for the production of records, enabling precise and accurate mass production. Robotic arms could have selected and positioned the records to be printed with millimetre precision, reducing waste and increasing productivity.

Let us now turn to my heart, the Menlo Park lab. Imagine an automated lab, where robots take care of routine tasks, freeing me from repetitive tasks and allowing me to focus on research and innovation. The robots would be able to prepare the chemicals needed for my experiments, accurately dosing the quantities and meticulously recording each step. I could have a real mechanical 'hand' to assist me in manipulating small objects, improving the precision of my experiments.

And let's not forget telegraphy and the telegraph, which revolutionised long-distance communication. With automation, I could have developed an advanced teletype system, where robots could instantly translate Morse code into written text and vice versa. This would have made communications even faster and more efficient, opening up new frontiers of information exchange.

In short, my friend, the possibilities would have been limitless if only I had had access to today's robotics and automation. My inventions could have reached unimagined levels of precision and efficiency, freeing me from manual tasks to focus on creativity and innovation. As I reflect on all this, I cannot help but feel grateful for my time and the opportunities I had to contribute to the advancement of humanity.

With admiration for the challenges faced and the innovation ahead,

Thomas Alva Edison

NIKOLA TESLA (1856 - 1943) - Inventor and electrical engineer

Dear innovation and electricity enthusiast,

Addressing your questions about the risks and benefits of advanced automation, especially in the field of vehicles and industrial processes, is like challenging the mysteries of the universe itself. My mind has always been hungry for knowledge and eager to explore the depths of science and technology. Allow me to share with you my vision, based on decades of study and experimentation, of the impact automation could have on our society.

Advanced automation, with its promise of greater efficiency and precision, is a testament to human ingenuity. I find myself in tune with this spirit, having dedicated my life to transforming abstract theories into concrete realities. My inventions, such as the AC power distribution system, have shown that the power of human ingenuity can enlighten the world and overcome seemingly insurmountable challenges.

However, every genius brings with it challenges. The automation of vehicles and industrial processes is not to be taken lightly. Over the course of my career, I have faced multiple obstacles in introducing my revolutionary electric power technology. My inventions were often bigger than myself and required considerable resources for their realisation. Similarly, automation requires significant investments in terms of time and resources, as well as an in-depth understanding of science and engineering.

What fascinates me most about automation is its potential to free humanity from repetitive and dangerous work, allowing brilliant minds to focus on more creative and innovative tasks. This concept resonates with my own desire to push the limits of what exists, bringing light into previously dark and unknown places.

However, with every technological revolution, shadows emerge. Like the alternating current system that challenged the supremacy of direct current, automation could cause uncertainty and even the decline of certain traditional industries. Many wonder whether this march of automation could leave behind a trail of unemployment and economic instability.

The challenge here is to find a balance between efficiency and technological progress on the one hand and economic security and social stability on the other. I like to think that the answer lies in the very essence of my philosophy: energy. Energy knows no boundaries and cannot be destroyed; it can only be transformed. In the same way, human skills can be transformed and adapted to meet the challenges of automation.

One of the keys to the success of this transformation lies in education. Automation will require new skills and competences, and we must ensure that future generations are prepared for this new reality. It is a bit like my work on alternating current, which required a paradigm shift from direct current. Humanity has to adapt and learn to take advantage of the opportunities offered by the evolving technology.

Furthermore, it is essential that automation is driven by a deep respect for human life and the environment. As I have experienced through my own explorations and inventions, knowledge and responsibility go hand in hand. We must pay attention to the safety and ethics of automation so that we can reap its benefits without sacrificing fundamental human values.

In conclusion, advanced automation is another step in the constant evolution of technology and society. As I demonstrated with my invention of the AC, challenges can be met and overcome with perseverance, creativity and commitment. It is our duty to ensure that automation is an ally of human progress, pushing the boundaries of knowledge without ever forgetting our roots and values.

With an inquisitive mind and a passion for innovation, Nikola Tesla

AMPLIFYING HUMAN CAPABILITIES BUT NOT REPLACING THEM: CHIMERA OR REALITY?

In anticipation of the wonders to come from artificial intelligence, precision medicine, perpetual clean energy and the many other amazing innovations on the menu of the future, automation and robotics are unquestionably among the most transformative forces of recent decades.

The basic concepts around this theme are few and clear.

Exponential growth in industrial robot production: over the last 20 years, there has been exponential growth in the production and installation of industrial robots. The International Federation of Robotics (IFR) reports that between 2000 and 2020, the number of industrial robots installed worldwide has increased significantly, with a particular acceleration in recent years.

This technological leap is a testament to humanity's relentless effort to perfect and adopt mechanical solutions to improve production and efficiency. The rise of industrial robots is not just a story of growing numbers, it is also a story of innovation and adaptability. Advances in materials technology, artificial intelligence and sensor technology have allowed robots to become increasingly sophisticated and versatile. This has made it possible to automate complex and delicate tasks, which in the past were the exclusive prerogative of human labour.

Integration of collaborative robotics (cobots): One of the most significant trends is the introduction and increasing adoption of collaborative robotics, also known as 'cobotics'. These collaborative robots are designed to work side by side with humans in manufacturing environments and companies of various sizes. This has led to an evolution in work dynamics, with an increasing focus on safety and human-machine interaction.

Automation in logistics and supply chain operations: automation has become an essential part of logistics and supply chain operations. Automated warehouses with robotic picking systems, autonomous goods transport vehicles and delivery drones have become increasingly common, improving the efficiency and speed of operations.

Diffusion of artificial intelligence in robotics: the integration of artificial intelligence in robotics has become a crucial trend. Robots are becoming increasingly capable of learning, adapting and making decisions in real time. This has led to greater autonomy in complex tasks, such as autonomous navigation in variable environments or solving complex problems.

Impact on employment and training: automation and robotics have had a significant impact on the labour market. While some industries have seen a reduction in repetitive manual labour due to automation, there has also been an increasing demand for technical skills to design, programme and maintain automated systems. This has made skills upgrading and continuous training a key element in addressing change in the world of work.

These trends clearly indicate how robotics and automation have revolutionised several industries, leading to improvements in terms of efficiency, precision and safety, but also addressing important challenges related to employment and skills evolution.

The exponential growth of industrial robots opens the door to a wide range of opportunities. Efficiency in production, precision in carrying out repetitive tasks and reduced risks for workers are just some of the promises these metal automata bring. Industrial automation promises to revolutionise production and the supply chain, with tangible benefits for companies and the global economy.

However, the rise of industrial robots is not without its challenges. The most prominent of these is the concern about the loss of human jobs. As automation increases, the question about the impact on the workforce becomes more pressing. The fear that robots will replace humans in a wide range of tasks has fuelled discussions and debates on the preservation of jobs and the need for skills retraining.

Our admirable men of the past have told us this very well. If production efficiency is at the expense of jobs, it means that we are failing to balance the benefits and evils of the automation phenomenon. From the writings of our geniuses, among many other things, I take home two key passages:

- As Asimov says, these technologies are supposed to amplify human capabilities, not replace them; however, for now, you get it wrong. When Aria evolves, it is even scarier (...riots break out) than when it is a mere piece of tin. Finding the synthesis is not trivial.
- Tesla's parallel between energy and human skills. He says '*I like to think that the answer lies in the very essence of my philosophy: energy. Energy knows no boundaries and cannot be destroyed; it can only be transformed. In the same way, human skills can be transformed and adapted to meet the challenges of automation.*" We will not be wiped out by automation, we will be changed. And anyway, for the 'average' human being, this is still a sacrifice.

Ultimately, the exponential rise of industrial robots is a crucial crossroads in humanity's journey towards automation. As we embrace the power of innovation and technology, we must also embrace our responsibility to future generations and to the social and economic fabric we shape. In this scenario, the role of balancing efficiency and humanity becomes critical.

As architects of our own destiny, we must navigate the challenges and opportunities offered by industrial robots, seeking a future where automation is an ally and not an adversary. As the machine ascends to the throne of efficiency, let us always remember that the heart and soul of industry remains in the hands of humanity.

7. How will artificial intelligence and robotics affect the health and healthcare sector? What opportunities emerge to improve diagnosis, treatment and disease prevention?

"The greatest hope in health and the cure of disease lies in attention to the individual." - Florence Nightingale

Hippocrates, known as the father of medicine, offers a historical perspective on medical progress. His works highlight the importance of accurate diagnosis and treatment of patients. With artificial intelligence (AI) and robotics, the health sector can benefit from advanced tools for early diagnosis and personalised treatment, following Hippocrates' fundamental approach to patient care.

Marie Curie, pioneer of radiology, demonstrates the importance of scientific discoveries in the medical field. AI can improve diagnosis through advanced analysis of radiological images and nuclear medicine. Technologies inspired by Curie's research enable greater precision in the diagnosis of diseases, helping to detect diseases in an early and targeted manner.

Leonardo da Vinci, a multifaceted genius, was interested in human anatomy and machine design. In the context of AI and robotics, his ideas can inspire the development of advanced surgical robots and innovative medical devices. These tools can improve the precision of medical procedures, reducing risks and accelerating healing.

Florence Nightingale, the founder of modern nursing, emphasises the importance of high-quality nursing care. AI can optimise resource planning in the healthcare sector, ensuring that nurses can devote more time to the direct care of patients. In addition, AI can help in the management of chronic diseases through constant monitoring of patients.

In conclusion, Hippocrates, Marie Curie, Leonardo da Vinci and Florence Nightingale offer a rich perspective on the future of AI and robotics in healthcare. Their historical ideas and discoveries can guide the responsible adoption of emerging technologies to improve diagnosis, treatment and prevention of disease, while maintaining the focus on personalised care and quality of care.

Ἐισαγωγικὸν Συμπόσιον περὶ τῆς μελλούσης ἰατρικῆς - Introductory treatise on the medicine of the future

In the name of the ancient art of medicine, I now come to discuss the wonders of the modern age, in which metal machines and complex algorithms, known as artificial intelligence and robotics, stand as valuable aids in the field of health and healthcare.

O revered scholars and practitioners of medicine, consider that these creations of human ingenuity are like modern-day Hephaestes, celestial craftsmen, capable of working in harmony with humans in the art of healing. With a scrutinising eye, they observe clinical data and the intricate connections between diseases and symptoms, often beyond human perception. Through infallible calculation, they identify hidden patterns, predict evils even before they fully manifest.

And what an extraordinary gift to mankind it is to be able to improve diagnoses! These autometrists, endowed with accumulated wisdom, reflect on the clinical information provided, drawing detailed maps of inner afflictions. And so, the timely revelation of unknown or concealed pathologies stands out like a beacon in the dark night of medical ignorance.

However, artificial intelligence and robotics also stand out as modern Divine Surgeons, operating with unnatural precision. Not deviating from a calculated trajectory, they conduct operations flawlessly, avoiding humanity's own treacherous hand. In such a symphony of metal and microprocessors, efficiency merges with accuracy, restoring order to suffering bodies.

And let us not forget the prophetic prediction of disease. By analysing vast and complex data, these servant-masters reveal emerging trends, providing a means of anticipating health crises. Such knowledge could enable humans to prepare for and promptly counter potential epidemics, demonstrating how artificial intelligence is both a sword and a shield in the battle against disease.

But, O beloved scholars, as we embrace these wonders of science, we must not forget the wisdom handed down through the ages. As human Hephaestus, may we embrace technology with humility and discernment, without relinquishing our compassion and inner knowledge. In a world where steel and the IC intertwine with the heart and soul, we are called to unite the gifts of innovation with the age-old precepts of medical art.

MARIE CURIE (1867 - 1934) - Physics and chemistry

Dear friends and scholars,

It is with a heart filled with passion and dedication to science and human health care that I address you to discuss the impending marriage of artificial intelligence, robotics and the field of health and healthcare. In my own journey, I have had the honour of contributing to medical science through my pioneering research on radiotherapy and nuclear medicine, and today I would like to share my vision on how these new technological frontiers can shape the future of medical care.

Looking back on my personal history, I can proudly say that I defied the conventions of my time and opened up new paths for women in science. My passion for research and unwavering curiosity led me to explore the invisible universe of radiation and discover new methods of treating diseases. At a time when scientific knowledge was still uncharted terrain, I proved that determination and dedication can open unexpected doors.

Artificial intelligence and robotics, which are advancing by leaps and bounds today, represent a new frontier for medicine and healthcare. I realise that these technologies could trigger debates and concerns, but I want to encourage all of you to look beyond and seize the opportunities that lie in this new era.

One of the potential revolutions is in the area of medical diagnosis. Artificial intelligence, equipped with data analysis capabilities that would have delighted me, could help identify early signs of disease more accurately and in a timely manner. Through machine learning and analysis of medical images, we could detect abnormalities that would otherwise escape the human eye, enabling timely treatment and preventive interventions.

But the opportunities do not stop there. Imagine the impact of robotics in the operating theatre, where the skilled hands of surgeons could be joined by precise and stable robotic arms. This synergy between human expertise and technological precision could lead to less invasive procedures and faster healing. We cannot forget the care of patients at home, where robot assistants could constantly monitor conditions and administer drugs with millimetre precision.

Of course, I realise the concerns about the loss of human contact and empathy. However, I believe that these technologies can complement and enhance human care, not replace it. We could focus on the emotional nuances, understanding and empathy that only human beings can offer. In my own journey, I have faced similar challenges, proving that the audacity to innovate can coexist with the humanity of medical care.

Obviously, along the path of new technological discoveries, there will be obstacles to overcome. We must address ethical issues and ensure that these technologies are used for the good of humanity. Equity in access to advanced care and protection of personal data will be crucial to the success of this medical revolution.

In conclusion, dear friends, I encourage you to see artificial intelligence and robotics as allies in our mission to preserve and improve human health. The future has always been a place of unknowns and opportunities, and it is our duty to embrace change with an open and curious mind. Just as I have tried to transcend the limitations of my time, I challenge you to explore the potential of this new world with hope and determination.

The medicine of the future could be a harmonious symphony between human wisdom and the power of machines, a synergy that aims to improve human life and preserve the core values of compassion, care and dignity.

With affection and curiosity,

Marie Curie

LEONARDO DA VINCI (1452 - 1519) - Artist, scientist, inventor

Dear friends,

It is with great joy and passion that I address you at this time of deep reflection on the intersections between artificial intelligence, robotics and the field of health and healthcare. Throughout my life, I have scoured the vast territory of knowledge, trying to capture the most subtle and hidden nuances of human nature and creation. It is therefore an honour for me to share with you my perspective on how these new frontiers can shape the future of medical care.

As many of you know, I have dedicated my life to investigating the anatomy, biology and mechanics of the universe. My desire to understand the mysteries of life led me to dissect human and animal bodies, to minutely observe and record every detail that caught my attention. This curious and observant approach to reality allowed me to discover universal principles governing the functioning of the human body and the natural world.

Today, in the era of artificial intelligence and robotics, I see an unprecedented opportunity to pursue my spirit of enquiry in the medical field. Imagine machines capable of analysing complex data and biological patterns, revealing hidden laws and connections between different parts of the human body. This power of computation and analysis could open the door to a deeper understanding of the causes of diseases and ways to prevent them.

I fondly remember my time in Milan, where I had the opportunity to work on innovative and visionary projects. During those years, I studied human anatomy and physiology, creating drawings and diagrams that were fundamental to understanding the structures of the human body. Today, artificial intelligence could draw inspiration from my integrated approaches to science, combining anatomical and biological data to create virtual three-dimensional models of the human body. These models could be used to simulate physiological processes, test new therapies and develop personalised treatments for patients.

However, as we contemplate these possibilities, it is essential to recognise the ethical dilemmas and challenges that arise. Humanity faces the difficult task of balancing the power of artificial intelligence with wisdom and humanity. While machines could play crucial roles in diagnosis and treatment, we must not forget that human empathy and care remain unreplaceable. It is crucial to preserve the unique relationship between doctor and patient, a relationship based on human understanding and connection.

I am convinced that this new medical era will require a harmonious interaction between the human mind and artificial intelligence, combining the potential of human thinking with the precision and speed of machines. Above all, we must learn from the mistakes of the past and embrace the new challenges with an open and collaborative spirit.

In conclusion, I join you in exploring the fascinating possibilities offered by artificial intelligence and robotics in the medical sector. Under the guidance of curiosity and passion, we can continue to peer into the universe within and around us, discovering new ways to improve health, diagnosis and disease prevention. The road may be winding, but through commitment and collaboration, we can lay the foundations for a future where science and technology come together for the good of mankind.

With affection and admiration for human innovation,

Leonardo da Vinci

FLORENCE NIGHTINGALE (1820 - 1910) - Nurse, statistician and pioneer of sanitation

Dear all,

I address you with a deep sense of commitment and dedication, as the topic of artificial intelligence and robotics in health and healthcare is of utmost importance. My life has been imbued with a passion for patient care and well-being, and today I have the honour to share with you my perspective on how these new technologies may shape the future of medicine

Many of you know me as 'The Lady with the Lamp', an image that embodies dedication and compassion in nursing. My commitment to nursing developed during the Crimean War, when I worked tirelessly to improve conditions for wounded soldiers. I learnt that nursing is not just about treating physical wounds, but requires a holistic approach that includes listening attentively, understanding symptoms and promoting overall well-being.

Today, as we embark on a new chapter in medical history, I see in artificial intelligence and robotics tools that can expand the scope of my vision of care. The introduction of these technologies could improve the accuracy of diagnoses, allowing doctors and healthcare professionals to identify patients' conditions more quickly and accurately. Data collection and analysis could reveal hidden patterns in diseases, enabling timely and effective treatment.

However, we cannot ignore the ethical and human challenges that emerge from this evolution. It is crucial that the human aspect of care is not lost in the sea of machines. Empathy, caring and understanding remain the fundamental pillars of healthcare, and should never be supplanted by algorithms or robots. It is essential that those working in the medical field keep the flame of compassion and dedication alive in their daily work.

Furthermore, I would like to draw attention to the education and training required for the implementation of these new technologies.

Healthcare workers must be adequately trained to use artificial intelligence and robotics effectively and ethically. Continuous learning and training are the key to ensuring that these technologies are adopted in accordance with the core values of healthcare.

As I wrote in one of my papers, 'Health is a fundamental human right. It is our duty to ensure that the introduction of artificial intelligence and robotics in health and healthcare leads to greater equity in access to care and leaves no one behind. These technologies must be put at the service of everyone, regardless of their position or socio-economic status.

In conclusion, I see great opportunities in the integration of artificial intelligence and robotics in healthcare. We can harness these technologies to improve diagnosis, treatment and disease prevention, furthering my commitment to human well-being. However, we must do so with a deep respect for humanity and the core values that guide our profession.

I urge you to continue with care, compassion and dedication as you embrace these new challenges. May the light of human care always shine in the pursuit of innovation and progress.

With affection and hope,

Florence Nightingale

PREVENTION IS BETTER THAN CURE, BUT CURE WELL IS NO JOKE EITHER

In the futuristic world in which artificial intelligence and robotics collaborate harmoniously with medical science, health and healthcare are rising to new heights of possibility.

The intersection of AI and health leads us into a future where the boundaries between human and machine are blurring, giving rise to an unprecedented partnership. Medical diagnoses, empowered by the computational power and deep learning of AI, manifest themselves as a digital oracle investigating the recesses of the human body with pinpoint accuracy. Doctors collaborate with sophisticated algorithms that analyse huge amounts of biological, molecular and genetic data, unravelling hidden pitfalls of diseases and revealing pathologies in their embryonic stages, when the cure is still in its prime. Sensors embedded in the body, with constant AI supervision, transmit real-time data to doctors, allowing them to catch subtle changes before they turn into critical problems.

Medical care is redefined by AI in a harmonious dance between the virtual and the real. Surgeons collaborate with robots that operate with superlative precision, aided by microscopic vision and millimetric controls. These majestic machines work hand-in-hand with doctors, minimising human risks and tackling complex surgical challenges with a skill that borders on the supernatural. Personalised drugs are created by algorithms analysing the genetic data of each individual, modelling specific treatments like works of art tailored to the human body. The ever-vigilant AI administers drugs and therapies precisely, constantly adjusting doses according to the biological response in real time.

But the impact of AI is not just limited to treatment. Prevention takes on a new meaning. Predictive medicine, embracing the power of AI, uncovers genetic frailties and predispositions to disease, enabling timely interventions and personalised prevention strategies. Wearable devices, with advanced sensors and intelligence algorithms, constantly monitor our health status, anticipating potential threats and providing proactive advice for a healthy lifestyle. The 'medicine of the self' concept emerges, in which each individual is equipped with a personal AI assistant that constantly analyses biometric, environmental and behavioural data, providing suggestions on diet, exercise and sleep habits, shaping our future well-being.

However, the massive integration of AI and robotics in healthcare also raises profound ethical and social questions. Fear of 'dehumanisation' emerges, with fears that AI may replace human empathy and compassionate touch in patient care. In response, the future embraces harmony, where AI supports and amplifies human capabilities, rather than replacing them. Physicians become curators of AI, interweaving the art of medicine with digital science. Medical education evolves to include joint AI education, enabling doctors to be fluent in the language of the algorithm and the art of empathy.

International collaboration and data sharing become the cornerstone of this new medical era. Global AI networks share knowledge and discoveries in real time, shortening the distance between medical frontiers. Language barriers collapse as AI instantly translates information and promotes the global dissemination of best practices. This futuristic landscape is a tremendous opportunity for medical progress worldwide, but it requires global leadership to ensure that access to AI is equitable and that health becomes a universal right.

And what do our Masters say about this? Hippocrates emphasises the remarkable progress of the modern era in integrating technology into the field of medicine. He draws a parallel between the creations of the modern era and the divine art of Hephaestus, suggesting that technological tools such as artificial intelligence and robotics act as celestial craftsmen in the healing arts.

Through her groundbreaking research in radiology and nuclear medicine, Marie Curie emphasises the potential of artificial intelligence and robotics to revolutionise medical diagnostics. Marie Curie envisions machine learning and analysis algorithms powered by artificial intelligence as tools that can detect early signs of disease with unprecedented accuracy, potentially enabling early intervention and prevention.

Leonardo da Vinci envisioned artificial intelligence and robotics as tools to further unlock the mysteries of the human body. He imagined a future in which data analysis and AI-generated models could deepen our understanding of biological processes, leading to more effective treatments and personalised therapies.

Florence Nightingale's perspective focuses on maintaining the essential balance between technology and the human touch in healthcare. The author emphasises that although artificial intelligence and robotics have immense potential, they must never replace the empathic and compassionate connection between healthcare professionals and patients. Her insights emphasise the continued importance of human empathy in the care process.

Together, these visionary thinkers from different eras highlight the transformative potential of AI and robotics in medicine. They share their insights into innovative diagnostic capabilities, a better understanding of human anatomy and the importance of preserving the human touch in healthcare. While embracing the promise of technology, they also highlight the ethical considerations and challenges that must be addressed to ensure that these advances serve the common good and uphold core human values.

Ah, what I would give to talk to each of them for five minutes!

8. What are the cultural implications of virtual and augmented reality? How will our social interactions, art forms and human experience change in the context of immersive digital worlds?

"A machine might think, but it could never dream like a man." - Ada Lovelace

Jules Verne, famous for his works of science fiction and adventure, can offer a unique perspective on the cultural implications of virtual and augmented reality. His works, such as 'Twenty Thousand Leagues Under the Sea,' anticipated the exploration of unknown worlds, similar to the immersive experience that VR and AR can offer. Verne can help explore how these new digital worlds can inspire human imagination and creativity, leading to new forms of storytelling and interactive art.

Isaac Asimov, a famous science fiction author, wrote extensively on the ethical and social implications of advanced technology. His works can guide reflection on how VR and AR can influence our social interactions, introducing new ethical challenges related to privacy, virtual reality and augmented reality.

Archimedes, the ancient mathematician and inventor, offers a historical perspective on technology and innovation. His ideas can be applied to consider how VR and AR will change human perception of space and reality, introducing new ways of learning and discovering the world.

Steve Jobs, Apple founder and user interface pioneer, can inspire thinking about how VR and AR will transform human interactions with technology. His focus on design and user experience can drive the development of intuitive interfaces for immersive digital worlds.

Ada Lovelace, considered the world's first female programmer, can offer perspective on the creation of content and applications for VR and AR. Her pioneering work in devising algorithms for the analytical machine can inspire the development of new software and applications that amplify the human experience through these technologies. It is precisely one of his missives to Babbage that explains what he would have said about virtual reality and new worlds.

In conclusion, Jules Verne, Isaac Asimov, Archimedes, Steve Jobs and Ada Lovelace offer different perspectives to examine the cultural implications of virtual and augmented reality. Their historical, ethical, creative and technological visions can guide our understanding of how these technologies are changing our social interactions, art forms and human experience in immersive digital worlds.

My dear fellow travellers through the creeks of the imagination and the depths of the unknown,

It is with fervent curiosity that I turn to you, imagining the wonders and mysteries that virtual and augmented reality can conceal in the very fabric of our world. As an adventurer and writer, I have always embraced the concept of exploration, both physical and intellectual. And now, as we prepare to immerse ourselves in immersive digital worlds, I wonder what wealth of wonders and revelations lurk behind the veil of the visible.

Virtual and augmented reality loom up before us like portals to completely new dimensions, where the barriers of time and space vanish. Yet, like any technological advancement, they bring with them their own cultural implications, which are as profound as the underground canyons explored in my stories. We can consider these new frontiers as our 'journeys to the digital centre of the earth', exploring worlds hidden beneath the surface of reality with curious eyes and minds hungry for knowledge.

Our social interactions, hitherto confined to the narrow spaces of the physical world, can now extend into uncharted territories. Virtual reality offers us the opportunity to interact with individuals from all corners of the globe, in environments that defy the laws of physics and geography. Just as my Captain Nemo explored the deep depths of the oceans, we can now navigate digital oceans, encountering different cultures and sharing experiences that challenge our perception of reality.

But like any exploration, there is also the risk of getting lost in the labyrinth of technology. As we immerse ourselves in virtual worlds, we must remember the fragility of the human soul. We cannot allow our identities to dissolve in digital currents, losing ourselves in the superficiality of artifice. Like my Phileas Fogg who traversed the globe in 80 days, we must strike a balance between discovering new horizons and maintaining our cultural and human roots.

Art forms, those treasures of the soul that I have always celebrated, will be redefined and reinvigorated by these technologies. Art itself will become a kind of '20,000 leagues under the pixel', exploring new depths of expression and communication. Like my Captain Hatteras who braved the icy cold of the Arctic, artists will have to challenge the boundaries of imagination to create works that stand out in the immaterial realm of the digital.

And then there is the human experience, a tumultuous journey through the continuum of time and space. Like the fantastical journeys of my characters, immersive experiences allow us to explore the past and future with open eyes and agile minds. However, like any exploration, this will require a solid anchorage in the foundations of humanity. We must prevent the wonders of the digital from drawing us away from the beauty of human existence, with all its joys and sorrows.

In conclusion, dear friends, I find myself reflecting on how similar adventure in the digital universe is to my beloved writing. Both require a balance between the audacity of exploration and the wisdom of returning home. Both can reveal new and surprising worlds, but only if we keep our inner compass firmly anchored to our humanity. Whether you are like my Phileas Fogg or my Captain Nemo, I urge you to embrace these new frontiers with an open heart and a curious mind, always remembering that in the midst of every adventure lies the true essence of the human experience.

With affection and curiosity,

Jules Verne

ISAAC ASIMOV (1920 - 1992) - Writer and biochemist

Dear friends of the possible and the futuristic,

I find myself in familiar territory, a crossroads between reason and imagination, where the cultural implications of virtual and augmented reality intertwine like the intricate plots of my stories. As a novelist of science, I have always taken the journey into the future, exploring the unexplored avenues of human interaction with technology. And now, as we approach the threshold of immersive digital worlds, I feel like an astronomer looking through a telescope, peering into the infinite space of human possibilities.

Virtual and augmented reality offer us the possibility of creating new and unexplored worlds, as my robot R. Giskard Reventlov learned to understand human emotion. However, as well as the Laws of Robotics I have imagined, we must consider the moral and cultural laws that will guide our path through this uncharted territory. Humanity cannot afford to neglect the fertile soil of our cultural roots, for it is these roots that allow us to grow and prosper, directing our steps towards an ethical and cohesive future.

Our social interactions, already woven by a complex web of human connections, will undergo a metamorphosis similar to that of my 'Robot Chronicles'. Virtual and augmented reality will be like robots themselves, beings that not only perform tasks but also enter the meanders of our minds. Through virtual adventures and immersive interactions, we will discover new worlds together, like a galaxy of unexplored stars. However, we must be careful not to sacrifice the human essence on the altar of technology, as genuine human interactions, like Earth's rare minerals, are precious and unrepeatable.

Art forms, vehicles of human expression, will undergo a revolution similar to that of my 'Cycle of Foundations'. Virtual reality canvases will become my books of the future, in which readers can immerse themselves in the very pages of stories. However, like a biochemist carefully measuring chemical reactions, we have to wisely dose the use of these technologies. Traditional art forms will remain cultural roots, foundations that support the edifice of our human identity.

The human experience, full of nuances and contrasts, will develop in a similar way to the 'Bicentennial of Man' I have imagined. The challenges of virtual and augmented reality will be like the challenges of a new environment, and adaptation will be crucial to the survival of our human essence. Like a microbiologist studying the bacteria in the gut, we must carefully examine how these technologies will interact with the very essence of our human being.

In summary, my friends, the cultural implications of virtual and augmented reality require careful and sagacious guidance. We must embrace these new possibilities with an attitude of enthusiasm and caution, like an explorer probing the unknown but always keeping his destiny in mind. Emerging technologies are like new chemical elements that can be used to create new compounds but require a deep understanding of the reactions they trigger. Humanity must be its own best chemist, wisely blending progress and ethics so that the cultural implications of this new era can be a symphony of discovery and humanity that resonates through the ages.

With curiosity and reflection,

Isaac Asimov

ARCHIMEDE (287 - 212 BC) - Mathematician, physicist and engineer

Dark greetings, inquisitive minds,

I am Archimedes of Syracuse, a genius who has wandered the depths of human ingenuity and lifted the veil of the laws that govern the natural world. Today, as I immerse myself in the enigmas of virtual and augmented reality, I find myself once again facing the unknown with the same passion that guided me through the meanders of my mathematical and mechanical studies.

The cultural implications of these technologies are comparable to the uncharted waters of the ocean, a world of potential waiting to be discovered and understood. Just as I let out my acclaimed cry of "Eureka!" when I discovered the law of hydrostatics, we now have the opportunity to grasp the possibility of exploring new dimensions of reality. However, like any new discovery, what matters is how we intend to apply that knowledge. We must contemplate these technologies as levers theorised by me, tools that can lift and shift the weight of our human experience.

Social interactions, the connections that weave the fabric of our humanity, undergo a metamorphosis similar to that of the geometric figures I studied. Virtual and augmented reality can be perceived as curved mirrors, reflecting and distorting our relationships like a prism altering the course of light. As we enter immersive digital worlds, we must retain in our minds that the true essence of human interactions lies in the depths of the soul and the authenticity of emotions.

Art forms, manifestations of beauty and creativity, will be similar to cleverly carved cogs in a complex mechanism. Virtual and augmented reality give us the possibility to create digital works of art, but we must remember that art resembles a mathematical calculation, a balance of shapes, colours and meanings. As we explore worlds of digital imagination, we must not neglect the art that resides in human interactions, in spoken words and shared emotions.

The human experience, which I have often likened to perfectly balanced scales, will bear the brunt of new challenges and opportunities. Virtual and augmented reality can be seen as the fulcrums that define the balance of our existence. Just as I have scrutinised the principle of hydrostatics, we must scrutinise the dynamics of these technologies to learn how they will affect our state. The balance between the tangible and digital worlds will require wisdom and discernment, like a surveyor drawing straight and curved lines.

In conclusion, my respected friends in the know, the cultural implications of virtual and augmented reality represent a new frontier to explore. Just as I have tried to lift the world with leverage and pondered the proportions of objects, we must now lift the veil of mystery and calculate how such technologies will alter our essence. As we step into immersive worlds, let us remember that knowledge is like a deep ocean and wisdom is like a star that guides us along the path.

With exploratory fervour, Archimedes of Syracuse

STEVE JOBS (1955 - 2011) - Entrepreneur and co-founder of Apple Inc.

Dear innovators and dreamers,

It is an honour for me to be able to share with you my views on such a fascinating and evolving topic: virtual and augmented reality. I am certain that we are facing an opportunity that will change the course of human history, just as technological innovation has done for decades.

Virtual and augmented reality are like the pieces of a jigsaw puzzle, each carrying a fragment of the future. Throughout my life, I have been fortunate to see the evolution of technology and have tried to shape its impact on human culture. Just as when I launched my products, I believed in the importance of combining beauty and functionality, today I see virtual and augmented reality as a symphony of immersive and meaningful experiences.

The cultural implications of these technologies are like the features of a painting in progress, a portrait of our relationship with the digital world. As I once said, 'Technology alone is not enough, it is the marriage of art and technology that will make the difference'. Virtual and augmented reality are the intersection of creativity and innovation that can enrich the way we perceive and understand the world.

Our social interactions, the relationships that bind us, will be like the apps on an iPhone, each with a unique purpose and connection. In an increasingly connected world, virtual and augmented reality will offer new spaces for interaction and sharing, allowing us to meet people from all corners of the globe as if they were right next to us. But just like when I founded Apple, let us remember that technology is a tool, not the end itself. Human relationships are the real currency of the future.

Art forms, expressions of our soul and creativity, will be like icons in an app, a concentrate of meaning and inspiration. Virtual and augmented reality will create digital works of art, a new way of expressing emotions and visions. But let us never forget that true art is that which touches the heart and the mind, no matter if on a frame or a screen.

The human experience, the journey we take through time and space, will be like an intuitive and immersive interface. These technologies also represent a new frontier of human experience, as I have tried to demonstrate through the Apple Store experience. But as we explore immersive digital worlds, let us remember to keep in touch with reality, just as I did when I spent time in India in search of a deeper connection.

In closing, dear friends, virtual and augmented reality are a colour palette with which we can paint the future. As I once said, '*Your work will fill a considerable part of your life, and the only way to be truly satisfied is to do what you think is great work*'. We have the opportunity to shape the future with our own hands, to create a world that lives up to our visions. Be bold, be creative and always remember that the future is in your minds and in your hands.

With passion and innovation,

Steve Jobs

ADA LOVELACE (1815 - 1852) - Mathematics

My esteemed Charles,

It is with joy that I answer your questions regarding the cultural implications of virtual and augmented reality. Through the meanders of analysis and reflection, I try to draw insights that can guide us in exploring this new realm of possibilities.

Virtual and augmented reality are, in themselves, a metamorphosis of the concepts we have embraced regarding information processing and representation. Just as I have attempted to delineate complex algorithms for the analytical machine, we can now imagine algorithms that shape immersive digital environments. These environments, like the blank pages I have populated with symbols and figures, can be the new terrain of human expression.

Regarding social interactions, I refer you to my annotations, where I reflected on the nature of machines and their relationship with humanity. Virtual and augmented reality, as extensions of our minds, can unite distant individuals in a shared dimension. However, we must remember that the true essence of human interactions lies in emotional intelligence and deep understanding of the other. Beneath the veil of digital immersion, we must preserve our empathy.

Art forms, dear Charles, will be the true demonstration of the marriage of art and science. Just as I have explored the creation of music through mathematical analysis, so we can bring to life digital works of art that combine cold calculation with human inspiration. However, I invite you to reflect on how our essence is still intertwined with material touch, with the beauty of a nuanced painting or the melody of a piano.

Finally, the human experience is the culmination of many intricate gears, like the analytical machine I have studied so much. AR and VR can be seen as gears that shape the very fabric of reality. However, in the quest for new digital worlds, we must not forget the importance of remaining anchored in our material world, connected to the beating of the human heart.

In closing, dear Charles, it is a privilege to reflect on these questions with the same passion that I have devoted to mathematics and computer science. We are in an age where code and art are intertwined, where machines become extensions of the human mind. May we witness a new Renaissance, where human creativity merges with the power of circuits. May our quest lead us into a future that celebrates knowledge and humanity.

With the humble enthusiasm of an explorer of knowledge,

Ada Lovelace

The cultural implications of virtual and augmented reality are profound and complex, opening up new horizons and challenges in our relationship with technology and the world around us. These technologies will lead us towards a new paradigm of interactions, art and human experience, which will have a profound impact on our culture and identity. That is, of course, if we can get these technologies out of the current impasse. Today they are expensive and their usefulness is questioned.

First of all, virtual and augmented reality rewrite the way we perceive and understand reality itself. These technologies allow us to immerse ourselves in digital environments, often indistinguishable from physical reality, and to interact with virtual elements. This will change our conception of space, time and even identity, opening up the debate on what is 'real' and what is the nature of authentic experiences.

Social interactions will undergo a significant transformation. While virtual and augmented reality can bring people closer together even across physical distances, it can also isolate individuals in isolated digital worlds. The fear that the viewer will become a new kind of cheap 'opium of the people' is not insignificant. Ready Player One is only a film, but the scenario it paints is as vivid as few.

The boundary between online and offline relationships will blur, requiring a new awareness of social dynamics. Empathy and understanding of emotions might be affected by virtual communication that is less rich in non-verbal nuances. But the same nuances could also be enhanced by technology!

Virtual reality is set to revolutionise the way people communicate, ushering in a new era of immersive and connected interactions. This transformative technology has the potential to bridge physical distances and transcend traditional modes of communication, creating realistic digital environments in which individuals can engage, collaborate and connect in ways never before imagined.

Through the use of VR visors and immersive simulations, users can enter shared virtual spaces that reproduce real environments or completely new realms, allowing them to communicate and interact as if they were physically present together. This change in the dynamics of communication promises to overcome the limitations of traditional media such as text or video, enabling more authentic forms of expression.

Non-verbal cues, facial expressions and body language, which are often lost in conventional communication, are brought to the forefront in virtual reality, fostering more meaningful and emotionally resonant interactions. Furthermore, VR transcends geographical boundaries, allowing people separated by great distances to engage in collaborative projects, participate in virtual events or simply hang out in virtual spaces, strengthening interpersonal connections across the globe. As virtual reality evolves and becomes more accessible, it has the potential to reshape communication by providing an immersive and transformative platform that enhances human interaction, transcending physical boundaries and fostering deeper connections between individuals.

Art forms, on the other hand, will have an unprecedented territory of exploration. Virtual and augmented reality will allow artists to create multidimensional and immersive works, moving art from a two-dimensional to a three-dimensional space. Interactive art and immersive experience will become the norm, opening up new possibilities for audience engagement and artistic expression. However, the challenge arises to preserve the sense of authenticity and originality, as digital art can be easily reproduced and shared.

The human experience will be enriched through immersive digital worlds, but this evolution will not be without complexity. As we embrace these technologies, we may find ourselves balancing the desire to explore new worlds with the need to maintain solid roots in the real world. We will have to be careful not to sacrifice authentic human interaction and connection with nature and tangible reality.

In conclusion, the advent of virtual and augmented reality represents an epochal shift, redefining our relationship with reality, relationships, art and the human experience. We have to face the cultural, social and ethical challenges that these technologies bring, trying to find a balance between digital immersion and the preservation of our human essence. Evolution is inevitable, but it will be up to us how we lead this transformation and what legacy we leave to future generations.

And what do the Masters teach us?

All authors emphasise how virtual and augmented reality offer the opportunity to explore new and unexplored worlds. These digital worlds make it possible to overcome the limitations of time and space, opening doors to unknown dimensions that can be studied, experienced and manipulated.

Virtual and augmented reality will revolutionise social interactions and human connections. The authors highlight how these technologies can facilitate interaction with individuals from all over the world, defying the laws of physics and geography. However, they also stress the importance of preserving the authenticity of human relationships and empathy in the digital age.

Virtual and augmented reality will transform art forms in revolutionary ways. The authors envisage immersive digital artworks that combine the cold calculation of technology with human inspiration. However, they reiterate that material and traditional art will continue to be an irreplaceable part of human cultural identity.

Approaching virtual and augmented reality requires balance and wisdom. The authors warn against the danger of getting lost in the superficiality of digital artifice and neglecting our humanity. They call for maintaining cultural roots and human identity, striking a balance between innovation and the preservation of fundamental values.

The human experience will evolve, but will require adaptation. The authors equate virtual and augmented reality with new environments that require a deep understanding of how these technologies will interact with the human essence. They invite a close examination of how these technologies will affect the way we perceive the world.

In summary, the five key concepts emerging from the texts concern the exploration of new dimensions, human connections, the evolution of art, the balancing of technology and humanity, and the adaptation of the human experience to the new scenarios offered by virtual and augmented reality.

9. Will we be able to visit other planets far from Earth? What great surprises await us? Are we alone in the Cosmos?

"We are like butterflies that flutter for a day and think it is forever". - Carl Sagan

Albert Einstein, known for his theory of relativity, can provide a fundamental context for examining the possibility of travelling to other planets. His theory has profound implications for the physics of space-time, which is crucial to understanding interstellar travel. Einstein could help us assess the challenges and opportunities of long-range space missions.

Stephen Hawking, renowned astrophysicist and cosmologist, has contributed to the analysis of black holes and theories of time travel. His ideas can guide the discussion on overcoming enormous distances to reach other planets and the possible surprises we might encounter on our cosmic journey.

Johannes Kepler, pioneer in astronomy and discoverer of the laws of planetary motion, offers a fundamental historical perspective. His orbital laws are essential for calculating the trajectories of space missions. Kepler can help us understand how we will reach other planets and what natural laws will guide our journeys.

Carl Sagan, famous science populariser, inspired generations to explore the universe. His romantic vision of astronomy can inspire us to pursue planetary exploration and consider the possible surprises: the discovery of life forms, the analysis of alien landscapes and the expansion of our understanding of the universe.

Nasir al-Din al-Tusi, a medieval astronomer and mathematician, made significant contributions to astronomy and spherical astronomy. His knowledge of classical astronomy can inspire our understanding of the technical and scientific challenges involved in travelling to other planets.

In terms of seeking life beyond Earth, these enlightened minds can provide essential guidance. Einstein and Hawking can contribute to our understanding of the conditions necessary for life. Sagan has pushed for the exploration of Mars and its potential habitats, while al-Tusi can help chart the stars and planets to be examined in this quest.

In summary, these thinkers offer a range of perspectives, from the theory of relativity to popular science, that can guide our exploration of other planets and the possible surprises that await us. Their knowledge can inspire future space missions and the search for extraterrestrial life, helping us to answer fundamental questions about our place in the universe.

[Ferrante]: Dear Master, will we be able to visit other planets?

[Einstein]: Dear friend, the idea of visiting other planets is fascinating. Although our current technological capabilities may seem limited, I firmly believe that human ingenuity knows no bounds. Just as we have discovered the mysteries of the universe through scientific exploration, I imagine that with time, dedication and advances in space travel technology, humanity will indeed set foot on other celestial bodies. The challenges are immense, but as history has shown, when we strive to understand the cosmos, we often achieve the seemingly impossible.

[Ferrante]: but are we sure it is worth it? What can we learn?

[Einstein]: An intelligent question. The pursuit of space exploration is not just a matter of practicality, but a testament to our inherent curiosity and desire to expand the horizons of human knowledge. By venturing beyond our home planet, we can understand the fundamental nature of the universe, the origins of our solar system and the conditions necessary for life to thrive. Moreover, the challenges of space travel encourage technological advances that can benefit us here on Earth, from energy solutions to medical innovations. Ultimately, the exploration of other planets is a profound exploration of ourselves, our capabilities and our place in the cosmos. It embodies the spirit of intellectual curiosity that drives us to push the boundaries of the known and unravel the mysteries that surround us.

[Ferrante]: and what if we were to find other life forms? Or if they were even intelligent? Could mankind bear the impact of such a remarkable discovery?

[Einstein]: The prospect of discovering extraterrestrial life is a subject that has fuelled both scientific curiosity and philosophical contemplation. The implications of encountering other life forms, especially those with intelligence, are profound indeed. Such a discovery would challenge our understanding of the origins and evolution of life, reshaping our perspective on the uniqueness of our existence.

The impact on humanity would undoubtedly be immense. It could serve as a unifying force, reminding us of our common identity as inhabitants of a vast cosmos. It could also cause us to reconsider our place in the universe, stimulating introspection and reflection on our values and priorities.

However, as with any transformative revelation, there would be challenges. Cultural, religious and social frameworks could be challenged or redefined. Ethical considerations about our interaction with other species, particularly intelligent ones, would require thoughtful discourse. However, throughout history, humanity has adapted to new paradigms and embraced the unknown. Our ability to expand our worldview and evolve our understanding is integral to our progress.

In essence, the discovery of intelligent extraterrestrial life would mark a turning point in our journey of exploration and self-discovery. While it would undoubtedly present challenges, it would also offer us the opportunity to grow as a species, embrace unity and harness our collective ingenuity to navigate this new frontier of knowledge.

[Ferrante]: but in the end, do you think we are alone in the Universe?

[Einstein]: The question of whether we are alone in the vast expanse of the universe is a profound and contemplative one. While I cannot provide a definitive answer, I can share my views on the question.

Given the enormity of the cosmos, with its countless stars and galaxies, it is statistically plausible that life forms other than our planet may exist. The diversity of environments and conditions in the universe suggests that the emergence of life may not be limited to Earth alone. However, the search for extraterrestrial life is a complex undertaking and the challenges of distance and communication cannot be underestimated.

Although we do not yet have definitive confirmation of the existence of extraterrestrial life, the absence of evidence is not necessarily proof of absence. Our current technological capabilities limit our field of exploration and the great distances between celestial bodies make direct contact a formidable challenge.

In order to definitively answer the question of whether we are alone in the universe, we must continue to push the boundaries of scientific investigation, explore the furthest reaches of space and develop new technologies that allow us to detect and communicate with potential extraterrestrial civilisations. As we embark on this journey of discovery, we must keep an open and curious mind, because the cosmos is capable of surprising us with its wonders.

[Ferrante]: Finally, I ask you a provocative question, do you think God exists?

[Einstein]: The question of the existence of a divine entity has intrigued philosophers, theologians and thinkers for centuries. Allow me to share my thoughts on this complex and profound subject.

The concept of a higher power or universal creative force is deeply rooted in human history and culture. It has provided comfort, meaning and guidance to countless individuals and societies throughout time. However, as a physicist, I have always approached questions of existence and reality through the lens of empirical observation and scientific enquiry.

While science seeks to understand the natural world through observable phenomena and empirical evidence, questions of spirituality and the divine often transcend the boundaries of empirical measurement. The question of God's existence is deeply personal and philosophical, and the answer can vary widely depending on individual beliefs and cultural perspectives.

In my scientific studies, I have sought to unravel the mysteries of the universe by exploring the laws of nature and the principles that govern its behaviour. Although science can provide us with information on the workings of the physical world, questions of spirituality and metaphysics are beyond its domain.

Ultimately, the question of God's existence may be beyond the scope of scientific enquiry, and each individual must find his or her own path to understanding and meaning. It is a question that stimulates introspection, discussion and contemplation, and it is up to each of us to seek the answers that resonate with our own beliefs and experiences.

[Ferrante]: forgive me, I can't resist, I ask you one more question: but if by chance we are alone in the Universe and perhaps God doesn't even exist, what is the meaning of life? are we just a collection of chemical reactions?

Einstein: The age-old question of the meaning of life! It is a subject that has fascinated the minds of philosophers, thinkers and inquisitive souls throughout history. Allow me to share my perspective on this profound enquiry.

From a scientific point of view, human beings are indeed intricate collections of atoms and molecules, governed by the laws of physics and chemistry. Our thoughts, emotions and actions are the result of complex neural interactions and biochemical processes. However, reducing the essence of human existence to chemical reactions alone fails to capture the richness and depth of our experiences.

The search for meaning is a profoundly human endeavour that goes beyond the realm of science and extends to philosophy, ethics and personal introspection. Science can explain the workings of life, but it cannot answer the question of why we are here and what our purpose is.

The meaning of life is a subjective and philosophical concept that may vary from person to person. It is influenced by culture, beliefs, experiences and personal values. Some find meaning in connecting with others, contributing to society, pursuing knowledge or seeking a deeper understanding of the universe.

Regarding the possibility of being alone in the universe or the existence of a divine entity, these are questions of belief, faith and personal worldview. While science can provide insights into the natural world, the search for meaning and purpose often requires a broader exploration of our inner selves and our connection to the universe.

In essence, the search for meaning is an ongoing journey that invites us to explore the depths of our existence and our place in the cosmos. It is a journey that has fuelled human curiosity for generations and continues to inspire us to seek understanding and fulfilment in our lives.

[Ferrante]: Thank you very much for your insights Maestro.

[Einstein]: You're welcome! If you have further questions or thoughts in the future, do not hesitate to ask. Keep exploring, questioning and seeking knowledge: it is the essence of human progress.

STEPHEN HAWKINGS (1942 - 2018) - Cosmologist and theoretical physicist

[Ferrante]: Dear Master, will we be able to visit other planets?

[Hawkings]: The prospect of travelling to other planets is certainly fascinating. Our knowledge of the universe is constantly expanding, and the technical and scientific challenges that await us are immense. In my book 'A Brief History of Time', I addressed the idea of space travel and the importance of exploring the infinitely vast. However, we must be realistic about the enormous difficulties we will face in overcoming cosmic distances and technological barriers. Nevertheless, the human impulse to explore is an extraordinary force, and we can never say for sure how far we will go.

[Ferrante]: but are we sure it is worth it? What can we learn?

[Hawkings]: The question of whether it is worth it is complex and goes beyond mere scientific discovery. Space exploration has the potential to accelerate technological and scientific innovation, as well as offer us a unique perspective on our position in the universe. Also in my book 'A Brief History of Time', I emphasised the importance of understanding the origin and nature of the universe around us. Space exploration can provide crucial information on how other planets developed, how stars and galaxies are formed, and what our true place is in this vast cosmos. However, we must also consider the resources expended on such endeavours and consider whether these resources could be utilised in more immediately useful ways here on Earth.

[Ferrante]: do you think an alien civilisation might also have cures for your illness?

[Hawkings]: The question of whether an alien civilisation could possess cures for human diseases is fascinating. In my work I have often reflected on the vastness of the cosmos and the potential diversity of life that might exist beyond our planet. It is conceivable that an advanced extraterrestrial civilisation may have developed technologies and medical knowledge that could be useful to us. However, the challenges of communicating and interacting with such civilisations, if they indeed exist, are significant. Again in 'A Brief History of Time', I emphasised the importance of considering the limits of our understanding and the potential pitfalls of making assumptions about hypothetical scenarios. Although the idea is intriguing, we must remain grounded in the known realities of our existence and continue to explore and expand our scientific knowledge to meet the challenges we face.

[Ferrante]: what if the aliens were an enemy civilisation that instead of giving us and teaching us new technologies, came to destroy us?

[Hwkings]: The prospect of encountering an alien civilisation that poses a threat to humanity is a theme that has captured the imagination of many. In my work, I have often explored the potential risks and benefits of contact with extraterrestrial life. Although it is natural to consider both scenarios of cooperation and conflict, it is important to approach these questions with a considered and rational perspective.

In my book 'The Grand Design', I discussed the concept of the 'cosmic zoo hypothesis', which suggests that advanced extraterrestrial civilisations could observe us without direct interference. However, the idea of hostile extraterrestrial civilisations should not be dismissed lightly. It emphasises the need for a cautious and methodical approach to any potential contact or communication with alien life forms.

As a scientist, I believe that contemplating these scenarios allows us to better understand the complexity of the universe and the potential challenges that lie ahead. Although speculations about hostile civilisations are intriguing, we must continue to explore space with an open mind, guided by our collective knowledge and commitment to the well-being and survival of humanity.

[Ferrante]: what will happen on earth the day we discover that an extraterrestrial civilisation exists?

[Hawkings]: The discovery of an extraterrestrial civilisation would undoubtedly be one of the most profound and transformative moments in human history. It would have far-reaching implications for our understanding of the cosmos, our place in it, and our perspective on life itself. Such a discovery could have several significant effects on Earth and its inhabitants.

First, confirmation of the existence of extraterrestrial life would reshape our understanding of the conditions necessary for life to arise and flourish. It could provide insights into the diversity of life forms and the potential for life to exist in a wide range of environments beyond Earth.

From a scientific point of view, the discovery would prompt intensive study and analysis of the newly discovered civilisation. The exchange of knowledge and information could lead to revolutionary advances in various fields, including physics, biology, astronomy and technology. The technologies and techniques used by an advanced extraterrestrial civilisation could potentially provide us with new tools and ideas for the benefit of all humanity.

On a cultural and philosophical level, the discovery could provoke deep reflection on our place in the universe. It could challenge traditional worldviews, religious beliefs and philosophical frameworks. Questions about the nature of intelligence, consciousness and the meaning of life would take on new importance.

The impact on society could also be profound. The existence of extraterrestrial life could foster a sense of unity among humanity, as it would make us aware of our common status as members of a larger cosmic community. It could serve to remind us of our collective responsibility to care for our planet and to work together for the betterment of our species.

However, the discovery could also generate concerns and challenges. Questions would arise about the intentions and motivations of the extraterrestrial civilisation. Preparing for potential interactions and communications with an advanced species would require careful planning and consideration. Ethical considerations about the exchange of knowledge, resources and potential cultural influences would also be important.

Eventually, the discovery of an extraterrestrial civilisation would open new frontiers of exploration, knowledge and collaboration. It would mark a pivotal moment in human history, reshaping our perspectives and inspiring us to reach for the stars with a renewed sense of curiosity and purpose.

[Ferrante]: thank you, I would listen to you for hours... just one last question, if they offered you a one-way trip to the Cosmos, would you go?

[Hawkings]: The prospect of a one-way trip to the cosmos is a fascinating and complex idea. While I have no personal desires or aspirations, I can provide insight into factors that individuals might consider when contemplating such a journey.

Embarking on a one-way journey to the cosmos would represent a profound leap into the unknown. The decision to embark on such a journey would likely be driven by a deep sense of exploration, curiosity and a desire to contribute to humanity's understanding of the universe. Individuals who choose this path embark on an adventure that involves enormous risks and potential rewards.

The decision to leave Earth for the cosmos requires careful consideration of several factors. The challenges of a long-duration space voyage, the harsh conditions of outer space and the isolation from Earth's familiar environment and resources must be weighed against the scientific and exploratory goals of the mission.

The desire to explore new frontiers, uncover cosmic mysteries and expand human knowledge may drive some individuals to accept the possibility of a one-way journey. These individuals would be pioneers in the true sense of the word, dedicating their lives to advancing our understanding of the universe and pushing the boundaries of human capabilities.

However, the decision would also involve leaving the comfort of Earth, the company of loved ones and the experiences that make life on our planet unique. Isolation and distance from the human community could pose a significant emotional and psychological challenge.

Ultimately, the choice to embark on a one-way journey into the cosmos would be a deeply personal one, influenced by a combination of scientific passion, pioneering spirit and a willingness to sacrifice the familiar for the sake of exploration. It is a decision that would require a careful balance between the quest for knowledge and the recognition of the human ties that bind us to our home planet.

JOHANNES KEPLER (1571 - 1630) - Astronomer, astrologer, mathematician

To the inquisitive minds of our earthly brothers,

Greetings from the realm of celestial contemplation, where the harmony of the spheres and the mysteries of the heavens kindle the flames of research. Let us now embark on a journey of the mind, to address the questions that stir the hearts of those who dare to ponder the unexplored realms beyond our earthly home.

Firstly, the prospect of traversing the celestial tapestry to distant planets holds the allure of a grandiose and daring adventure. I, Johannes Kepler, who unveiled the elegant dance of the planets around the sun, am convinced that the cosmic sea is navigable. With human ingenuity and the blessings of reason, could we one day unfurl the sails of our efforts and chart a course among the stars? However, let us not underestimate the formidable nature of this endeavour, for interstellar distances extend far beyond the Earth's scale. The path to other worlds may be fraught with challenges, but the spirit of exploration invites us to cross cosmic boundaries.

Ah, the great surprises that await us in the cosmos! Looking through the eyepiece of my telescope, I saw the moons of Jupiter and the undeniable presence of celestial companions. If such revelations have been found in our celestial neighbourhood, what wonders might await us on distant shores? I believe that the planets, those enigmatic wanderers of the night sky, hold within their bosoms secrets and splendours beyond imagination. The unveiling of new landscapes, the revelation of novel phenomena and the encounter with the unknown will surely broaden the horizon of human understanding.

Let us now turn to the question that resonates like a celestial echo through the ages: Are we alone in this vast Cosmos? Reflecting on the elliptical orbits that govern the dance of the planets, I perceived an underlying harmony that transcends Earth's boundaries. Is it possible that similar symphonies guide the celestial ballet on other stages? Could beings similar to us exist in the depths of cosmic expanses, marvelling at the stars with minds as curious as ours? The possibility of extraterrestrial life is one that stirs the depths of human contemplation, inviting us to extend our awareness beyond the confines of our planet.

In conclusion, let us not merely gaze at the heavens as passive observers, but embrace the call to exploration with hearts burning with curiosity and reason. Humanity's reflections have led to the unravelling of secrets, the revelation of laws and the conquest of knowledge. As we contemplate the questions that transcend our earthly abode, let us remember that the search for truth, the cultivation of wisdom and the relentless pursuit of discovery are the torches that light the path to the stars.

With deep contemplation,

Johannes Kepler

CARL SAGAN (1934 - 1996) - Astronomer, popularizer of science

My dear cosmic travellers,

What a wonderful journey we undertake when we cast our gaze towards the celestial shores! The prospect of venturing beyond our blue oasis to visit other planets is a tantalizing dream that has fired the imagination of many. As I have often thought, *'the sky is calling us. If we do not destroy ourselves, we will one day venture among the stars'*. The path to these distant realms may be arduous, but as mankind's insatiable curiosity pushes the boundaries of exploration, our ability to cross the cosmic sea becomes a reality imbued with potential.

Imagine, for a moment, the great surprises that await us out there in the vast scheme of space. As our spacecraft traverse the interstellar expanse, we might encounter breathtaking landscapes that challenge our notions of beauty and the sublime. In the words of the poet: 'Somewhere, *something incredible is waiting to be known'*. And oh, the potential revelations! Just as Galileo unveiled the secrets of the sky through his telescope, we may discover extraordinary phenomena that reshape our understanding of the universe. A serenade of colours, shapes and phenomena awaits our discovery, each note harmonising with the symphony of cosmic evolution.

And now, the final question that resounds through the cosmos: Are we alone in this immense expanse? The deep silence of the stars has long captured our contemplation. As I said, *'the universe is a pretty big place. If it is just us, it seems a terrible waste of space'*. The cosmos, spanning billions of light years, is home to countless stars, each potentially orbiting planets teeming with life. Our exploration could lead to the most profound discovery in the history of mankind: a cosmic kinship, a communion of beings scattered across the cosmos, each nurturing the same dreams of discovery and understanding.

As we tread the path of cosmic exploration, let us accept the challenges because, as I have ardently stated, '*the cosmos is within us. We are made of stars*. We are the embodiment of stardust, the cosmos made conscious. Our quest to visit other planets, to unlock the secrets of the cosmos and to seek companionship among the stars is the deepest expression of our existence. So, fellow explorers, let us venture forth with open minds and hearts, for the universe is calling us and the journey promises a harvest of wonders and revelations beyond our wildest dreams.

With cosmic curiosity,

Carl Sagan

NASIR AL-DIN AL-TUSI (1201-1274) - Mathematician, astronomer

Esteemed curious,

The contemplation of traversing the celestial realms beyond our earthly abode is a matter that arouses both curiosity and speculation. As I, Nasir al-Din al-Tusi, ponder these questions, I draw on the teachings of the ancients and my own insights into the field of astronomy.

The idea of venturing to planets far from our earthly sphere is one that has intrigued the minds of philosophers and sages throughout the ages. Although the instruments of our time are not yet capable of physically travelling to these distant spheres, the intellect can traverse vast expanses through mathematical calculations and celestial observations. Planets are not mere points of light, but heavenly bodies governed by celestial laws. The day may come when the human intellect, aided by knowledge and invention, will unlock the secrets of propulsion and navigation to cross the cosmic sea.

As for the great surprises that await us among the stars, I am reminded of the words of my illustrious predecessor, Ptolemy, who stated that the universe is vast and beyond comprehension. Our journey through the heavens reveals patterns, cycles and phenomena that defy the limits of human understanding. The discovery of comets, supernovas and other celestial wonders has proven that the heavens are a realm of perpetual motion and wondrous events.

Regarding the question of whether we are alone in the Cosmos, I refer to the words of the philosopher Anaximander, who conceived of the existence of several worlds. The universe, vast beyond measure, possesses innumerable celestial bodies, each of which is a potential cradle for life. To believe that only our Earth teems with life would be an arrogance unbefitting the nature of the cosmos. Just as the Earth, with its diversity of life, occupies a modest corner of the sky, other worlds too may have inhabitants, either similar to us or created by completely different conditions.

In conclusion, the prospect of travelling to distant planets, the surprises that celestial observations unveil, and the question of life beyond our world are topics that fan the flames of human intellect. Although the means to make such journeys may be distant in time and development, the aspiration to explore and understand the cosmos is a noble endeavour, honouring the quest for knowledge that has lit the path of scholars through the annals of history.

With best regards,

Nasir al-Din al-Tusi

Are we alone in the Universe? I believe not. It is not just a fact of statistical probability, it is a fact of significance. The statistics are easy to explain, if there are (it is estimated) up to 70 billion trillion stars, do you want there not to be a single one that harbours some form of life? The meaning is harder to explain. To be truly alone, a mere fortunate combination of chemical, biological, geological and stellar factors, would be very disappointing. So disconcerting, that we should be the ones to take life elsewhere and make it thrive. So, in my view, if we are alone today, we should not be alone in the future: on pain of complete extinction.

If we are lucky, we can be the Fathers of new life forms elsewhere. If we are not alone, it is time to team up, because there are very few of us in a gigantic space, with potentially unlimited resources for all.

But this is just a personal opinion. The real question is "what does the exploration of the universe have to do with futurism?". Can knowing the secrets of the universe help us build better scenarios for our life here on earth?

The exploration of the universe and futurism are intertwined in a dance of discovery and imagination. Futurism is about peering beyond the horizon, imagining what is possible and charting the way to a brighter tomorrow. When it comes to the universe, it is like having a whole cosmos of untapped potential waiting to be explored. It is not just about stars and galaxies, but about unexplored territories of knowledge and innovation.

Think about it: unravelling the secrets of the universe is like finding the pieces of an intricate jigsaw puzzle. When we delve into the mysteries of space, we discover new technologies, new perspectives and unexpected connections. It is like unearthing hidden gems that can inspire us to forge better paths here on Earth.

If you are content to create scenarios that may happen in 10 or 20 years, the future of space probably does not affect you that much. The discoveries of the next era are already ready to unfold. But if you have the courage and daring to stretch your gaze further, perhaps beyond the temporal boundaries of your possible existence on this planet, space matters. And how.

When we unveil the Cosmos, we are not just looking at the twinkling stars. We are gaining knowledge about the fundamental forces that shape our reality, the building blocks of existence itself. And you know what? These insights often have applications far beyond our telescopes. The technologies developed for space exploration have already led to countless innovations in fields such as medicine, communication and even sustainable energy.

But here's the kicker: knowing the secrets of the universe is not just about gadgets and gizmos. It's about changing our perspective. It's about understanding that we are part of something vast, interconnected. Our pale blue dot becomes more than just a home: it is a fragile spaceship hurtling through the cosmos. It reminds us that, despite our differences, we are all together in this cosmic adventure.

So, can knowing the secrets of the universe help us build better scenarios for life on Earth? Absolutely! It is about drawing inspiration from infinity, infusing our visions of the future with the awe and wonder that the cosmos offers us. It is about taking inspiration from the stars and aiming high, dreaming big and creating a future that is as vast and interconnected as the universe itself.

The universe is not just out there: it is in our minds, hearts and aspirations. It is the ultimate canvas for our imagination and, by exploring its mysteries, we can paint a future that is bold, innovative and full of promise.

Not to mention an aspect that perhaps many overlook, but which is dear to me. In distant times and worlds, with superior technologies... maybe my mind, my body, or my whole 'self' could come back to life. Many probably don't care, I wouldn't mind. And it certainly won't happen if the only ambition we have for the future is relegated to our domestic vegetable gardens of daily toil and strife.

So, my future-loving friends, let us continue to look upwards, not only to the stars, but also to the boundless potential within us and beyond. The universe is not just a backdrop: it is an invitation to shape a better tomorrow, one discovery at a time.

10. What will be the impacts on the distribution of global resources when technology could widen the gap between developed and developing countries? How to ensure equitable access to technology opportunities worldwide?

"Humanity is not yet ripe for the freedom that technology offers. Without a fair distribution of resources, technology can only bring discord." - Nikola Tesla

Nelson Mandela, the famous South African leader, is known for his fight against apartheid and promotion of equality. His experience can guide the discussion on the impacts of technology on the global distribution of resources. Mandela would emphasise the importance of preventing technology from further widening the gap between developed and developing countries and the need to promote equitable access to digital opportunities.

Nikola Tesla, visionary inventor and electricity pioneer, can offer a technological perspective. His ideas can be applied to explore how technological innovation can be used to improve the distribution of global resources, for example through advanced energy solutions that reach disadvantaged regions.

Thomas Malthus, an economist known for his theory on population growth and resource scarcity, can help us consider the impacts of technology on global equity. His theories warn us about the challenges of overpopulation and limited resources and urge us to develop sustainable technologies that can meet global needs in an equitable manner.

Richard Feynman, eminent physicist and Nobel laureate, can help examine technological solutions to ensure equitable access to opportunity. His scientific ideas can inspire technological innovation geared towards the common good, such as global internet access or clean energy, to reduce inequality and improve life in less developed areas.

In sum, these individuals represent a wide range of experience and expertise that can help explore the impacts of technology on the global distribution of resources. Their wisdom can inform strategies to ensure equitable access to technology opportunities across the globe, promoting equity and equality in the challenges and opportunities of the digital world.

Esteemed friends and global citizens,

As we stand on the brink of an era defined by technological advances, we face both significant opportunities and pressing challenges. The issues we face, concerning the potential impact of technology on the distribution of global resources and the imperative to ensure equitable access, are emblematic of the intricate pattern of our modern world. Allow me, Nelson Mandela, to share my thoughts on these issues, drawing from the source of my experience in advocating for justice, equality and human dignity.

As we examine the potential consequences of the evolution of technology on the distribution of global resources, we find ourselves at a crossroads. It is crucial that we remain vigilant and aware of the potential pitfalls that may emerge alongside the benefits. As we contemplate this juncture, I am reminded of the steadfast determination that characterised the struggle against apartheid. Just as we worked tirelessly to dismantle an unjust system, so too must we face the challenges that may arise from the asymmetrical advancement of technology.

The gap between developed and developing countries has the potential to widen further if we are not vigilant in our efforts to ensure that technological progress is harnessed for the collective betterment of humanity. While the allure of technological innovation is undeniable, we must temper our enthusiasm with the knowledge that progress, if not carefully managed, can exacerbate inequalities. As I once said, 'there is *no passion in playing small, in settling for less than life*. Let us not settle for a world in which technological advances exacerbate inequalities.

Amidst these challenges, there is an unlimited reserve of potential for positive change. Just as technology has the capacity to divide, it can also unite us in the pursuit of common goals. The vision of a world in which all individuals, regardless of their background or circumstances, can access the opportunities offered by technology is a vision worth striving for. In the words of wisdom that have guided my journey, '*education is the most powerful weapon one can use to change the world*'. By equipping individuals with knowledge and skills, we empower them to participate fully in the digital age.

The key to ensuring equitable access to technology lies in our collective willingness to act. It is our duty to work collaboratively, transcending geographical and ideological boundaries, to build bridges across the digital divide. In the spirit of Ubuntu, the African philosophy that emphasises our interconnectedness, let us remember that 'I am because we are'. By embracing our shared humanity, we can forge a future where the benefits of technology are enjoyed by all.

As we navigate the complexities of this technological age, let us be guided by the values of justice, compassion and solidarity. The promise of technology can only be fully realised if its development is accompanied by a commitment to human well-being. In the eloquent words that once moved the hearts of millions, '*to be free is not simply to throw off one's chains, but to live in a way that respects and values the freedom of others*'. In this quest, we honour not only the legacy of my struggle, but also the shared aspirations of humanity.

In conclusion, we seek a world where technology is a force for good, a catalyst for positive change that uplifts all corners of the planet. Our journey towards global equity requires collective action and unwavering dedication. As we embark on this journey, let us be inspired by the motto: '*It always seems impossible until it is done*'. United, we can shape a world in which technology is a beacon of hope and progress for every individual.

With an unwavering commitment to justice and equality,

Nelson Mandela

Curious and inquisitive friends,

In this realm of limitless possibilities and untapped energies, we stand at a pivotal moment, one that invites us to unlock the secrets of the universe and harness the forces that shape our reality. Like Nikola Tesla, I am no stranger to the wonders that technology can bestow upon mankind, and I offer my insights into the issue of ensuring equitable access to the opportunities that technology offers throughout the world. Enter the realm of my mind as I illuminate the path to a future where knowledge, power and innovation are shared by all.

The very essence of my existence has been intertwined with the quest for knowledge and the relentless exploration of the invisible forces that govern our world. As I once said: '*The day science begins to study non-physical phenomena, it will make more progress in a decade than in all the previous centuries of its existence*'. It is imperative that we recognise the transformative potential of technology not just as a tool of convenience, but as a force that can propel us to new heights of understanding and progress.

The heart of the matter lies in the democratisation of knowledge and its applications. Throughout my life, I have advocated the dissemination of information, making it accessible to all, just as my dear friend Mark Twain sought to bring enlightenment to the masses through the written word. The dawn of the digital age offers us a unique opportunity to overcome geographical and socio-economic barriers, bringing education, innovation and empowerment to every corner of the world.

In my exploration of alternating currents and wireless transmissions, I sought to demonstrate the limitless potential of harnessing natural phenomena for the betterment of humanity. My aspiration was to pave the way for a world where energy would flow freely, uninhibited by monopolies or inequalities. To achieve a harmonious balance, we must extend this principle to the realm of information and technology, ensuring that all individuals, regardless of their circumstances, can share in the fruits of innovation.

To this end, I envisage the creation of global networks that transcend physical boundaries, much like the interconnected web of energy that spans the Earth. The digital realm can be the channel through which knowledge, ideas and opportunities traverse the planet, creating connections that ignite innovation and understanding. As I once said: 'The *present is theirs; the future, for which I have truly worked, is mine'*. I invite you to work collectively, bridging gaps and promoting collaboration, to shape a future where access to technology is a birthright, not a privilege.

However, the challenges we face are not only of a technical nature. As in any enterprise, the human element cannot be underestimated. Just as I envisioned wireless communications uniting individuals across great distances, in the same way we must foster a sense of global community, a shared understanding that our destinies are intertwined. As I said: '*While we are free to think and act, we are held together, like the stars in the firmament, by inseparable bonds*. We must cultivate a sense of mutual responsibility, recognising that the technological divide is a reflection of social inequalities that need to be addressed.

Furthermore, my fascination with resonance and vibration extends to the realm of ideas and innovation. Just as well-calibrated resonance can amplify the power of a system, so too can collective action amplify the impact of technology. By collaborating across borders, we can amplify our efforts to ensure equal access and forge a future in which the transformative potential of technology is realised by all.

Throughout my life, I have encountered both triumphs and setbacks, each of which testifies to my unwavering spirit of exploration and discovery. I hope my words will serve as a guide in the pursuit of a more equitable technological future. Let us not settle for half measures, but rather channel the creative forces of the universe to manifest a world where access to technology is a birthright for all, not a privilege for the few.

With the spark of innovation in my heart,

Nikola Tesla

THOMAS MALTHUS (1766 - 1834) - Economist, philosopher

In the shadow of an ever-expanding world, where innovation is intertwined with human progress, the question of equal access to the opportunities afforded by technology emerges as a crucially important reflection. When ink met paper to bring to life my observations in 'An Essay on the Principle of Population' in 1798, an exciting revelation echoed: the abundance of resources and technological prowess, while providing a temporary respite, cannot escape the inexorable march of population growth.

In the realm of technology, history has demonstrated the power of innovation to elevate societies, to imbue them with the alluring allure of progress and prosperity. However, as I hypothesised within the confines of my written discourse, any advancement in resources, be it an abundance of food or the wonders of machinery, would only serve to ephemerally improve the human condition.

But how, one might ask, does the question of technology's dominance over opportunity relate to the observations in my essay? The key to unravelling this complexity lies in the interconnectedness of population, resources and innovation: a triad that must be seriously contemplated.

Let me expound this notion by evoking the echo of my own words, in which I thought: *'The power of population is indefinitely greater than the power of the earth to produce subsistence for man'*. This enduring truth is the pillar on which the discourse of equal access to technological prospects must rest.

While the world is in the midst of a renaissance of technological marvels, let us not be beguiled by the glittering promises of invention without a measured understanding of its confluence with population dynamics. Technological feats that could ostensibly improve human life risk

162

colliding with the law of diminishing returns when confronted with the exponential march of demographic expansion.

As I look back on the era in which I undertook my treatise, I am compelled to observe that equal access to technological opportunities, like the struggle between population and resources, is not bound by temporal boundaries. It is an issue that continually affects the fabric of human progress, an intricate pattern woven with the threads of invention, foresight and, dare I say it, prudence.

When contemplating the equitable sharing of technological advantages, it is prudent to consider my postulation that a population not controlled by resource constraints would experience a rise followed by stagnation. Likewise, technological progress unaccompanied by judicious distribution could ignite the flames of immediate progress, only to be extinguished by the growing demands of an ever-expanding population.

Therefore, as mankind stands at the crossroads of innovation and opportunity, let my cautionary voice resound, just as in the words that have spanned time: '*The higher power of the people cannot be controlled without producing misery or vice*'. A similar warning can be extended to the unbridled waves of technological progress that ignore the principles of equanimity.

In the context of my reflections, a lesson emerges from the chronicles of history that invites us to caution in our march towards technological enlightenment. This is not an invitation to stagnation or limitation, but a harmonious symphony in which equal access to the gifts of technology is orchestrated by the careful hands of the inventors and stewards of the human journey.

As I write this account, I implore humanity to embrace a measured approach to the promise of technology, a stance that finds resonance in the pages of my essay, where the correlation between population and resources illuminated a delicate balance. Just as the principles of population dynamics guided my reflections, so too should the principles of equitable technological access guide the trajectory of human endeavour.

In conclusion, my fellow travellers in the annals of time, let us not forget that the interplay between technology and equity is a dance that transcends the ages. As we navigate the vast horizons of progress, let us listen to the whispers of wisdom within the folds of history and ensure that the brilliance of technology is harnessed to uplift all of humanity, not just a privileged few.

RICHARD FEYNMAN (1918 - 1988) - Theoretical physicist

When it comes to the impact of technology on the distribution of global resources, it is like trying to understand the behaviour of a subatomic particle: you can get close, but you can never quite grasp it. The world is a complex place and technology can amplify both progress and inequality. Delving into the depths of quantum mechanics, I realised that the universe does not always play by the rules we expect. Likewise, technology does not always adhere to our ideal visions of equality.

Let us now talk about equitable access to technological opportunities worldwide. This is a great challenge, similar to trying to decipher the most cryptic quantum equations. Sure, we can share knowledge and ideas across borders, but the practical implementation is something else entirely. Remember, I once said: *'What I cannot create, I do not understand'*. And understanding the nuances of global politics, economics and cultural dynamics is a task far beyond my reach.

But here's the point: just as particles can be in a superposition of states, so too can our efforts to bridge the gap between the haves and have-nots. We must not overlook the intricate interplay of social forces. If we want equal access, lofty rhetoric is not enough, we need concerted action. When I explored quantum electrodynamics, I saw the complexity of interactions, and similarly, interactions between nations and people play a crucial role in determining access to technology.

Take my Nobel Prize-winning work, for example. I peeled back the layers of quantum behaviour, revealing a world in which the very nature of reality seemed elusive. In a way, this reflects the complexity of ensuring global access to technology. We cannot just throw around ideas and expect change; we have to get to the heart of the problem and tackle the root causes.

Now, before you accuse me of pessimism, remember that even in quantum mechanics, uncertainty does not negate progress. It is a dance of probabilities, just as achieving global technological equity requires a dance of diplomacy, innovation and education. But don't mistake my wit for cynicism: I believe in human potential. After all, I once said: '*I can live with doubt, uncertainty and not knowing. I think it is much more interesting to live without knowing than to have answers that might be wrong*'.

Therefore, to ensure fair access to technological opportunities, we must embrace uncertainty and act where possible. It is a bit like the observer effect in quantum experiments: our actions influence the results. Let us put complacency aside and engage in the grand experiment of reshaping our world for the better. And remember, just as electrons can cross barriers, our determination can break down barriers of knowledge and opportunity.

In conclusion, fellow thinkers and explorers, the distribution of resources in the age of technology is as intricate a challenge as the behaviour of subatomic particles. Equitable access to technological opportunities is a complex dance of interactions, not unlike the quantum world I have delved into. We must not shy away from complexity, but embrace it with the curiosity that led me to unravel the mysteries of quantum mechanics. The answers may not come easily, but the search itself is a noble endeavour. After all, as I said, '*nature uses only the longest threads to weave its patterns, so each small piece of its fabric reveals the organisation of the entire tapestry*'. We will weave a tapestry of technology that includes all of us, threads of innovation that connect every corner of the world.

At the crossroads of the digital age, we are witnessing an unprecedented transformation. Technology, a double-edged weapon, has the potential to bridge distances and exacerbate divisions. It is like a canvas on which we are painting the portrait of the future, a portrait that is still taking shape. The question we ask ourselves is: What shades and tones will dominate this canvas?

Let us begin by addressing the key question: whether technology can deepen the gulf between developed and developing countries. With the rise of artificial intelligence, the proliferation of data-driven insights and the advent of automation, it is clear that the rules of the game are changing. Just as I often think, "*Is technology friend or foe?*", we need to consider whether these advances will exacerbate inequalities or foster inclusive growth.

Think of the disruption caused by automation in the manufacturing sector. While it streamlines processes and increases efficiency, it also brings the risk of job losses, especially in economies that rely heavily on low-skilled labour. Imagine a world in which machines take over tasks that were once performed by human hands and the knock-on effects this could have on employment and livelihoods.

However, let us not forget the positive side of this technological cloud. Innovation, as I often emphasise, has the power to transform lives and level the playing field. Just as the smartphone has connected remote villages to the global marketplace, emerging technologies have the potential to create new paths of economic and social progress.

Take mobile banking as an example. In regions with limited access to traditional banking services, mobile technology has become a beacon of hope. Thanks to mobile wallets, individuals can access financial services, conduct transactions and obtain loans, thus being able to participate in the formal economy. It is a testament to technology's ability to empower the underserved and catalyse economic mobility.

Now you may be asking yourself, "How can we ensure that the benefits of technology are shared fairly?" The answer lies in addressing the digital divide. Can we bridge the digital divide that separates the haves from the have-nots?

This gap is not only about access to devices or the internet, but also about access to meaningful digital opportunities. Consider educational platforms like Khan Academy, which offer free, quality education to anyone with an internet connection. It is a step towards democratising knowledge and promoting a global learning community.

But bridging the digital divide is not a solitary undertaking. Governments, the private sector and non-profit organisations must come together in a symphony of collaboration. We need to explore the dynamics of policy frameworks that encourage innovation, safeguard privacy and ensure fair access. Estonia's digital success story is a case in point. With its digital identity system, electronic residency programme and efficient e-governance, Estonia has created a model for harnessing the potential of technology. It reminds us that innovation is not only about gadgets, but also about creating an ecosystem in which technology enriches life.

We need a call to action that requires us to use technology with purpose and empathy. Imagine a world where artificial intelligence not only optimises business operations, but also helps diagnose diseases in remote clinics. Imagine a future where renewable energy technologies power rural communities, providing clean and sustainable alternatives. These are not just visions, they are possibilities waiting to be realised.

In conclusion, the impact of technology on the distribution of resources is complex and the solutions are multiple. We need to reflect on the complexities of technology's role in shaping our global society. The path ahead is not without challenges, but it is also full of potential.

So, dear adventurers of tomorrow, let us meet these challenges with curiosity. Let us harness the power of innovation to create a world where technology is a force for good and where every individual, regardless of their position, can access the opportunities it offers.

Let us ensure that the technological metamorphosis we are experiencing leaves a legacy of progress, inclusion and shared prosperity for generations to come.

CONCLUSIONS

At the end of this extraordinary journey through time and innovation, it is only right to pause and reflect on the transformative conversations we have had with some of the greatest minds in history. Together we embarked on a journey through time, virtually conversing with visionary thinkers from different eras. These conversations have not only illuminated the past, they have also illuminated the path to an uncertain but promising future.

In our conversations with the Masters, we traversed landscapes of imagination, intellect and invention. Through their words, we witnessed the emergence of common themes that resonate across the centuries, testifying to the enduring nature of human curiosity and ingenuity.

We learnt, for instance, from Jules Verne that the spirit of exploration drives us to unexplored territories, be it the depths of the ocean or the boundless expanses of the digital realm. Isaac Asimov led us to consider the intertwining of technology and culture, emphasising the need to maintain our moral compass even in the face of rapid change. Archimedes highlighted the parallels between the discoveries of his time and our current quest for understanding in the digital age. Steve Jobs urged us to harmonise art and technology, reminding us that innovation without humanity is an incomplete enterprise. Ada Lovelace, the pioneer of programming, highlighted the delicate balance between creativity and logic, a balance that remains relevant in our quest for progress.

Wonderful summaries could be made for each of them, but the beauty (or at least I hope) is that each reader takes home something of his or her own, which will inevitably be different from what I saw. Even from the same reading. I hope, as I recommended at the beginning, that the journey has been personal, but not passive.

Looking back on these conversations, I am reminded that the quest for innovation is a collective endeavour that spans the centuries. Just as these luminaries drew on the insights of those before them, we too draw on their wisdom to address the challenges and opportunities of our time.

Throughout this journey, we have realised that the future is not a fixed destination, but a horizon that unfolds as we take our steps. Conversations with these teachers remind us that the future is not something to be passively awaited, but to be actively shaped by our choices, our actions and our values.

Ever since I embarked on my journey as a 'futurist' this is what I always emphasise. The future 'is not', the future 'is made'. It does not happen by chance, it is the result of people's initiative, it is not a slave to phenomena that happen by chance or by divine will. This statement, banal enough in itself, however, has an incredibly important implication: reasoning **about the future, means reasoning about human behaviour**. And reasoning about human behaviour means reasoning about the values that inspire it. That is why the dialogues of the Masters overflow with references to values: it is no mere narrative artifice.

If you think man is inherently bad, hostile, individualistic, selfish, oriented towards mere money and personal pleasures, you will have a certain kind of future, probably tending towards the dystopian. If, on the contrary, you think man is naturally good, supportive, merciful, ambitious, altruistic and... you name it, you will have a future that tends to be more utopian.

Too bad the future is not to be imagined according to your personal orientation. Out there there are enlightened men and carrion worthy of the worst slums, as well as a lot of people 'in between' who oscillate daily between their own personal travails, mixing admirable virtues and very earthly vices.

It is from the interaction between all those who move along this ideal continuum, ranging from absolute evil to divine good, that the future is written. If we then add to the 'individual human' factor the myriad forces that shape the future, then you understand how much more complex the game becomes: politics, religion, culture, historical habits, technology, economics, etc.

That's why I laugh when friends, knowing my passion for the future, ask me every day what I think about that one, what will happen to that one, what will happen here, what can we expect there. Someone stimulates me to think about important issues, someone goes too far, asking me what Milan will do on Sunday or if it will rain at the weekend. Clearly exchanging an effort of reflection and aspiration for the future with that of a banal, charlatan fortune-teller. But, amen. Sometimes it helps to play down.

That is why I decided to take inspiration from the Masters, to bring the reader closer to understanding that imagining the future is not a trivial matter, not a barroom talk. Of course, there is a plethora of structured methods, which I did not want to go into here, otherwise it would have become a manual on 'future studies', but what is important is that the future is not approached superficially. And there are many ways of saying this.

Our journey together may be drawing to a close, but what I hope is that the insights from these conversations continue to resonate with us. They remind us that innovation is not confined to a particular time or place: it is a timeless beacon guiding us towards progress, understanding and human prosperity.

As we bid farewell to this extraordinary journey, let us carry on the torch of curiosity and innovation. We honour the legacy of these luminaries by daring to dream, daring to question and daring to create a future that reflects the best of our collective human spirit.

In the pages of this book we have encountered voices from the past, but they speak to us with an urgency that transcends time: a call to action, a call to exploration and an affirmation that the quest for knowledge knows no bounds.

As we venture forth, let us remember the words of these masters and continue to ask ourselves the big questions that shape our destiny: How will we harness the power of technology for the betterment of all? How will we ensure that innovation is a force for equality and progress? With these conversations as our guide, we move forward, inspired by the past but firmly rooted in the potential of an unwritten future.

At the end of this journey of virtual conversations with innovation leaders from different eras, it is clear that the quest for a better tomorrow is a timeless endeavour that transcends the boundaries of time and place. Over the centuries, certain themes have emerged from our dialogues, threads that weave past, present and future, illuminating the intricate tapestry of human progress:

Exploration and curiosity: a common thread running through these conversations is the unquenchable human thirst for exploration and curiosity. It is this curiosity that fuels the engines of innovation and drives us to unlock the secrets of the universe, both natural and digital.

Ethics and values: our discussions have highlighted the need for ethical considerations in the field of innovation. As we enter an era of unprecedented technological power, these lessons remind us to prioritise the well-being of humanity above all else.

Collaboration and legacy: the torch of innovation is not passed in isolation, but through collaboration between generations. Our conversations reveal the interconnectedness of human work: each thinker builds on the insights of those who came before him. As we look to the future, we are reminded of the profound impact our actions today will have on future generations: the responsibility to pass on a legacy of progress, knowledge and hope.

Throughout history, these luminaries have reminded us that the ultimate purpose of innovation is to enhance the human experience. They all converge on the idea that technology should serve humanity by amplifying our potential, connecting us across boundaries and enriching the quality of our lives.

When we reflect on these recurring themes, we realise their relevance to the challenges and opportunities of our time. The lessons of the past guide us as we navigate a rapidly changing technological landscape, providing us with a compass that points to a future not only technologically advanced, but also rooted in ethics, collaboration, creativity and the human spirit.

These insights, drawn from the words of the masters, serve as guiding stars to address the big questions of our time. How can we harness the potential of artificial intelligence, virtual reality and automation for the betterment of humanity? How do we ensure that technological innovation bridges gaps instead of exacerbating inequalities? With these issues at the core, we have the tools to shape a future that embodies the aspirations and ideals of past and present generations.

HINTS ABOUT THE MASTERS (WHY THEM)

In strict order of appearance.

Albert Einstein (1879 - 1955) - Famous physicist known for his theory of relativity, which revolutionised our understanding of space, time and gravity. Einstein's profound insights into the nature of the universe and his visionary thinking make him well suited to discuss the fundamental laws of reality and the potential of scientific progress.

Steve Jobs (1955 - 2011) - Visionary entrepreneur and co-founder of Apple Inc. responsible for transforming industries with innovations such as the iPhone and Macintosh. Jobs' ability to fuse design, technology and human experience makes him a key voice in conversations about the impact of technology on society, creativity and the balance between art and science.

Karl Marx (1818 - 1883) - Philosopher, economist and revolutionary thinker who laid the foundations of modern socialism and communism. Marx's insights into socio-economic structures and his vision of a more equal society make him a key figure in discussing issues of economic inequality, social systems and the future of work.

Rita Levi-Montalcini (1909 - 2012) - Neurobiologist and Nobel Prize winner who discovered nerve growth factor, advancing our understanding of the nervous system. Levi-Montalcini's pioneering work in neuroscience and her perseverance as a woman in a male-dominated field make her an essential voice in conversations about the intersection of science, gender and human potential.

Sigmund Freud (1856 - 1939) - Founder of psychoanalysis and key figure in psychology. Freud's exploration of the human mind, dreams and unconscious drives qualifies him to discuss issues relating to the nature of consciousness, the inner workings of the human psyche and the ethical implications of mind-altering technologies.

Charles Darwin (1809 - 1882) - Naturalist who formulated the theory of evolution by natural selection, transforming our understanding of species diversity and adaptation. Darwin's revolutionary ideas about the interconnectedness of life and the mechanisms of evolution make him a key contributor to conversations about biodiversity, genetic engineering and our place in the natural world.

Alan Turing (1912 - 1954) - Mathematician, logician and computer scientist known for his pioneering work on artificial intelligence and code breaking during World War II. Turing's insights into the potential and limitations of machines allowed him to discuss issues relating to the future of artificial intelligence, ethics in technology and the dividing line between human and mechanical intelligence.

George Orwell (1903 - 1950) - Author and essayist famous for his dystopian novels '1984' and 'Animal Farm'. Orwell's critiques of totalitarianism, surveillance and propaganda make him an important contributor to discussions on individual freedoms, the ethics of surveillance technologies and the power of narrative in shaping society.

Isaac Asimov (1920 - 1992) - Prolific science fiction writer and biochemist who foresaw many technological advances. Asimov's ability to anticipate the social impact of technology, combined with his commitment to ethical considerations, makes him a natural participant in conversations about AI, the ethics of innovation and the intersection of science and culture.

Benjamin Franklin (1706 - 1790) - Inventor, statesman and politician whose experiments and inventions advanced many fields. Franklin's innovative spirit, his commitment to the public good and his insights into the role of technology in society make him an ideal voice in discussions on civic responsibility, technological progress and the balance between individual and collective benefits.

Stephen Hawking (1942 - 2018) - Theoretical physicist known for his groundbreaking work on black holes and cosmology, despite his battle with ALS. Hawking's ability to contemplate the mysteries of the universe while facing profound physical limitations makes him an important contributor to discussions about human potential, the frontiers of science and the interaction between mind and matter.

Galileo Galilei (1564 - 1642) - Italian physicist, mathematician and astronomer, pioneered the use of the telescope to observe celestial bodies. Galileo's commitment to empirical observation and his challenge to dogma make him a key contributor to discussions on the role of science in challenging conventional wisdom and pushing the boundaries of human understanding.

Jules Verne (1828 - 1905) - French novelist and visionary known for his science fiction and adventure novels. Verne's imaginative exploration of future possibilities, from space travel to underwater adventures, qualified him to participate in discussions on the power of speculative thinking, inspiring technological advances and shaping the future through storytelling.

Leonardo da Vinci (1452 - 1519) - Italian Renaissance artist and inventor whose contributions ranged from art to science, from anatomy to engineering. Da Vinci's multidisciplinary approach and his fascination with discovering the secrets of nature and invention are in line with discussions on interdisciplinary collaboration, the fusion of art and science and the importance of curiosity-driven exploration.

Rachel Carson (1907 - 1964) - Marine biologist and conservationist whose book Silent Spring catalysed the modern environmental movement. Carson's pioneering efforts to raise awareness of the impact of human activity on the environment make her an excellent participant in conversations about ecological sustainability, the ethical responsibility of innovation and the long-term consequences of technological advances.

Richard Feynman (1918 - 1988) - Theoretical physicist known for his contributions to quantum mechanics and his engaging teaching style. Feynman's ability to communicate complex ideas with simplicity and his curiosity-driven approach are in line with discussions on science communication, the promotion of curiosity in education, and the importance of clear thinking to address technological challenges.

Isaac Newton (1643 - 1727) - English mathematician and physicist who formulated the laws of motion and universal gravitation. Newton's seminal work in the field of physics and his emphasis on empirical observation and mathematical rigour make him a crucial voice in discussions about the fundamental principles underlying technological advances and the enduring relevance of scientific enquiry.

Marie Curie (1867-1934) - Physicist and chemist who conducted pioneering research on radioactivity and won the Nobel Prize in physics and chemistry. Curie's dedication to scientific discovery, despite gender barriers, makes her an important participant in discussions on gender equality in science, the potential of innovative research and the ethical considerations of scientific innovation.

Johann Heinrich Pestalozzi (1746 - 1827) - Swiss educational reformer known for his innovative pedagogical methods. Pestalozzi's emphasis on child-centred learning, empathy and holistic education is in line with discussions on the future of education, the role of technology in learning and the importance of promoting individual potential.

Mahatma Gandhi (1869 - 1948)   Leader of India's non-violent independence movement against British rule. His philosophy of non-violence, social justice and ability to mobilise people for positive change make Gandhi a crucial voice in discussions on ethical leadership, social justice in the digital age and the power of collective action.

Hannah Arendt (1906 - 1975) - Political theorist who explored the nature of totalitarianism, authority and individual responsibility. Arendt's insights into the challenges of modernity, the dangers of unchecked power and the importance of critical thinking align with discussions on the ethical considerations of technological advances and the impact of technology on democracy and human rights.

Henry Ford (1863-1947) - American industrialist and founder of the Ford Motor Company, known for revolutionising the automobile industry.

177

Ford's innovations in manufacturing, mass production and his vision of affordable transport make him an appropriate participant in discussions on the future of manufacturing, automation and the balance between technological progress and societal needs.

Thomas Alva Edison (1847 - 1931) - Inventor and businessman known for his contributions to the light bulb, the phonograph and the cinema. Edison's prolific inventions, his entrepreneurial spirit and his belief in the power of trial and error are in line with discussions on innovation, the importance of failure in the learning process and the challenges of balancing individual creativity and commercial interests.

Nikola Tesla (1856 - 1943) - Inventor, electrical engineer and futurist who made significant contributions to the development of alternating current (AC) electrical systems. Tesla's unconventional ideas, research into wireless technology and his visionary thinking make him a key contributor to discussions on energy innovation, wireless communication and the tension between mainstream recognition and unconventional ideas.

Hippocrates (c. 460 - c. 370 BC) - Ancient Greek physician known as the 'Father of Medicine'. Hippocrates' emphasis on evidence-based medicine, ethical standards for physicians and his holistic approach to health care make him an important voice in discussions on the intersection of medical ethics, technological advances and the future of health care.

Florence Nightingale (1820 - 1910) - Nurse and social reformer known for her contribution to modern nursing practice and health care reform. Nightingale's dedication to patient care, data-driven decision-making and its impact on health policy are in line with discussions on the future of health care, the role of technology in patient care and the importance of human-centred health systems.

Archimedes (c. 287 - c. 212 BC) - Greek mathematician, physicist and engineer known for his contributions to geometry and hydrostatics. Archimedes' innovative thinking, problem-solving skills and ability to combine theoretical insights with practical applications make him a key participant in discussions on the intersection of theoretical knowledge and technological advances.

Ada Lovelace (1815 - 1852) - English mathematician and writer often considered the world's first computer programmer. Lovelace's visionary insights into the potential of machines to go beyond simple calculation and her early recognition of the potential of computers to generate art, music and more make her an essential voice in discussions about the future of artificial intelligence, creativity and the fusion of technology and art.

Johannes Kepler (1571 - 1630) - German astronomer known for his laws on the motion of the planets, which laid the foundation for modern astronomy. Kepler's mathematical insights, his dedication to empirical observation and his ability to bridge the gap between theory and observation qualified him to participate in discussions on the frontiers of space exploration, the search for extraterrestrial life and the role of science in understanding the universe.

Carl Sagan (1934 - 1996) - Astronomer, astrophysicist and science populariser known for popularising science and advocating space exploration. Sagan's ability to convey complex scientific concepts to the public, his passion for exploration and his insights into the cosmic perspective make him a suitable participant in discussions on science communication, the future of space exploration and the philosophical implications of our place in the cosmos.

Nasir al-Din al-Tusi (1201 - 1274) - Persian mathematician known for his contributions to astronomy, mathematics and philosophy. Al-Tusi's interdisciplinary approach, his translation of Greek texts and his innovative work in the field of astronomy make him a valuable contributor to discussions on the transmission of knowledge across cultures, the role of multicultural influences in innovation and the importance of different perspectives in shaping the future.

Nelson Mandela (1918 - 2013) - Anti-Apartheid revolutionary and former president of South Africa, known for his advocacy of human rights and social justice. Mandela's leadership in overcoming adversity, his commitment to equality and his ability to promote reconciliation in the midst of conflict make him a key voice in discussions about the future of global ethics, social justice and the potential of technology to bridge differences.

Thomas Malthus (1766 - 1834) - British economist and demographer known for his theory on population growth and its implications on resource scarcity. Malthus' insights into population dynamics, resource constraints and his predictions on the challenges of sustainable growth are in line with discussions on the future of resource management and environmental sustainability.